Mineral Resource Estimation

Mineral Resource Estimation

Editor

Anuradha Rawat

Mineral Resource Estimation

Edited by **Anuradha Rawat**

Printed in 2017

ISBN: 978-1-68117-148-7

Library of Congress Control Number: 2015936564

© 2016 by
SCITUS Academics LLC,
616, Corporate Way, Suite 2, 4766,
Valley Cottage, NY 10989

www.scitusacademics.com

Contents

Preface..vii

Chapter 1   Dissolution of Arsenic Minerals Mediated by Dissimilatory
            Arsenate Reducing Bacteria: Estimation of the Physiological
            Potential for Arsenic Mobilization...1
            Drewniak Lukasz, Rajpert Liwia, Mantur Aleksandra,
            and Sklodowska Aleksandra

Chapter 2   Land Use Suitability Assessment in Low-Slope Hilly Regions under
            the Impact of Urbanization in Yunnan, China....................................33
            Gui Jin, Zhaohua Li, Qiaowen Lin, Chenchen Shi, Bing Liu, and
            Lina Yao

Chapter 3   The Analysis of Pricing Power of Preponderant Metal Mineral
            Resources under the Perspective of Intergenerational Equity and
            Social Preferences: An Analytical Framework Based on Cournot
            Equilibrium Model 61
            Meirui Zhong, Anqi Zeng, Jianbai Huang, and Jinyu Chen

Chapter 4   Evaluation of Groundwater Recharge Estimates in a Partially
            Metamorphosed Sedimentary Basin in a Tropical Environment:
            Application of Natural Tracers 95
            Felix Oteng Mensah, Clement Alo, and Sandow Mark Yidana

Chapter 5   Forecasting the Development of Boreal Paludified Forests in
            Response to Climate Change: A Case Study Using Ontario
            Ecosite Classification..115
            Benoit Lafleur, Nicole J Fenton, and Yves Bergeron

Chapter 6   Priority Areas for Watershed Service Conservation in the
            Guapi-Macacu Region of Rio de Janeiro, Atlantic Forest, Brazil....143
            Vanesa Rodríguez Osuna, Jan Börner, Udo Nehren, Rachel Bardy
            Prado, Hartmut Gaese, and Jürgen Heinrich

Chapter 7   Potential Water-related Environmental Risks of Hydraulic
             Fracturing Employed in Exploration and Exploitation of
             Unconventional Natural Gas Reservoirs in Germany 193
             Axel Bergmann, Frank-Andreas Weber, H Georg Meiners, and
             Frank Müller

Chapter 8   Ecological Characterization of Soil-inhabiting and Hypolithic
             Soil Crusts within the Knersvlakte, South Africa 225
             Bettina Weber, Dirk CJ Wessels, Kirstin Deutschewitz, Stephanie
             Dojani, Hans Reichenberger, and Burkhard Büdel

             Citations 255
             Index 259

Preface

The process of estimating a Mineral Resource can only take place after the estimator is convinced of the soundness of the fundamentals underlying the estimation process. Thus the database of sampling, density, and other quality data for both estimation and geological interpretation must have integrity and robustness ; the geological data must be sufficiently complete for the definition of a geological model; the geological model itself must have internal consistency, should explain the observed arrangement of lithological and mineralogical domains, and should represent the estimator's best knowledge of the genesis of the mineral deposit; and the geological model should support the distribution of mineralisation seen in the sampling. It is only at this stage that a resource model may be generated.

Editor

Dissolution of Arsenic Minerals Mediated by Dissimilatory Arsenate Reducing Bacteria: Estimation of the Physiological Potential for Arsenic Mobilization

Drewniak Lukasz, Rajpert Liwia, Mantur Aleksandra, and Sklodowska Aleksandra

Laboratory of Environmental Pollution Analysis, Faculty of Biology, University of Warsaw, Miecznikowa 1, 02-096 Warsaw, Poland

ABSTRACT

The aim of this study was characterization of the isolated dissimilatory arsenate reducing bacteria in the context of their potential for arsenic removal from primary arsenic minerals through reductive dissolution. Four strains, Shewanella sp. OM1, Pseudomonas sp. OM2, Aeromonas sp. OM4, and Serratia sp. OM17, capable of anaerobic growth with As (V) reduction, were isolated from microbial mats from an ancient gold mine. All of the isolated strains: (i) produced siderophores that promote dissolution of minerals, (ii) were resistant to dissolved arsenic compounds, (iii) were able to use the dissolved arsenates as the terminal electron acceptor, and (iii) were able to use copper minerals containing arsenic minerals (e.g., enargite) as a respiratory substrate. Based on the results obtained in this study, we postulate that arsenic can be released from some As-bearing polymetallic minerals (such as copper ore concentrates or middlings) under reductive conditions by dissimilatory arsenate reducers in indirect processes.

INTRODUCTION

Arsenic is considered to be one of the most hazardous elements, wherein its toxicity is revealed only when it is present in aqueous or gaseous form. Most of the arsenic-bearing minerals, such as arsenides and sulfarsenides, are considered nontoxic because they are highly insoluble. Problems arise when these primary minerals break down and enter into solution or form more soluble species such as oxides [1]. One of such cases, a common phenomenon in metallurgy industry, is arsenic release from ores and deposits into the environment through mining and smelting operations.

The most abundant and dominant arsenic ore minerals are As-sulfides, including arsenopyrite (FeAsS), realgar (As_4S_4), and orpiment (As_2S_3) [2]. Arsenic minerals are also found as impurities (as minor ores) in the ores of other metals such as gold or copper. Arsenopyrite is very often found in gold ores, while enargite (Cu_3AsS_4) and tennantite (($Cu,Fe)_{12}As_4S_{13}$) are the most common arsenic-bearing minerals associated with copper sulfide ore bodies [3]. Moreover, the fine fraction of ash produced by smelting of ore concentrates causes the airborne dispersion of arsenic, thus contaminating soil and streams over a wide area. A high arsenic content causes problems in the smelting and further extraction of metals, resulting in a reduction in the quality of the final product [3–5]. Environmental problems are also connected with the arsenic-bearing flotation tailings and mine waters. Arsenic present in these "reservoirs" can be transformed by the microbial activity and this may contribute to the further dissemination of arsenic contamination.

The most common microbial arsenic mobilization processes involve oxidative dissolution of minerals through oxidation of iron, sulfur, or arsenic [6]. Iron oxidizers can promote oxidative dissolution of arsenic minerals through direct or indirect mechanisms, which results in the production of toxic arsenious (H_3AsO_3) and arsenic acids (H_3AsO_4) [7]. In turn, oxidation of arsenic-bearing sulfide minerals causes not only the release of arsenic but also the acidification of waters and their enrichment in sulfate anions and the accompanying heavy metals [8]. Arsenic minerals dissolution can be mediated also at circumneutral pH, but mainly by sulfur- and arsenite-oxidizing microbes. It has been documented that the release of pyrite-bound arsenic may be caused by (i) oxidation of sulfide in the pyrite lattice [9] or (ii) direct oxidation of arsenic from mineral structure (e.g., arsenopyrite) [10]. The release of arsenic from As-bearing minerals may be also mediated by microbial reductive dissolution. Under reducing conditions, microorganisms can use arsenic compounds as terminal electron acceptors in arsenic respiration [11, 12]. However, this process is thought to be dominant for arsenic displacement from secondary minerals, for example, arsenate adsorbed on scorodite [13] and ferrihydrite [14]. There is no data about dissimilatory reduction of primary arsenic minerals, such as enargite, fangite, or luzonite, in which arsenic occurs as (sulf) arsenates. Thus, the question is whether the microorganisms are capable of removing arsenic from primary arsenic minerals through reductive dissolution. The answer to this question will help to complete the missing knowledge about microbial dissolution of arsenic-bearing minerals and will help to estimate the chance of the potential use of dissimilatory reducers in biomining and bioremediation. The possibility of selective removal of As (III) compounds (whose solubility is higher than As (V) and the level of As (III) sorption on the surface of secondary minerals is much lower and unstable) will be particularly important in industrial processing of As-bearing polymetallic minerals (such as copper ore concentrates or middlings (intermediate during ore processing)), where the presence of arsenic causes many technological and environmental problems.

The aim of this study was characterization of the isolated dissimilatory arsenate reducing bacteria in the context of their potential for arsenic removal from primary arsenic minerals through reductive dissolution.

We demonstrated that the isolated strains (i) produce metabolites that promote dissolution of minerals, (ii) are resistant to dissolved arsenic compounds, (iii) are able to use the dissolved arsenates as terminal electron acceptor, and (iv) are able to remove arsenic under anaerobic conditions from As-containing copper ore concentrates and middlings, in which copper minerals were used as the sole terminal electron acceptor.

MATERIALS AND METHODS

Isolation of Arsenate Dissimilatory Reducing Bacteria, Media and Growth Conditions

Microbial mats samples were collected from an ancient gold mine located in Zloty Stok, Lower Silesia, SW Poland, and were used as the inoculum for isolation of dissimilatory arsenate reducers. Microbial mats and the mine waters are characterized by high arsenic content (~5000–6800 mg/L for mats and ~3000–7000 µg/L for mine waters), slightly alkaline pH (~7, 4–8.0), and stable temperature of 10–12°C throughout the year. 5 mL of microbial mats samples was added to the modified mineral salt medium (MSM) [10] (final volume, 100 mL) containing 2.5 mM sodium arsenate and 5 mM sodium lactate. Cultures were carried out in serum bottles with CO_2:N_2 (in a ratio 20: 80) injected into the headspace. These bottles were sealed and corked with silicon stoppers secured by aluminium crimp seals. Sampling was performed under CO_2:N_2 (in a ratio 20: 80) atmosphere in anaerobic glove box (Sigma-Aldrich). Cultures were incubated for 7 days at 22°C and then were subcultured twice for 7 days each. The cultures were then diluted and plated on MSM agar containing 2.5 mM sodium arsenate and 0.004% yeast extract.

All of the isolated strains were routinely grown in lysogeny broth (LB) medium [15] or in MSM medium [NaCl 1.17 (g/L), KCl 0.30 (g/L), NH_4Cl 0.15 (g/L), $MgCl_2.6H_2O$ 0.41 (g/L), $CaCl_2$ 0.11 (g/L), KH_2PO_4 0.20 (g/L), Na_2SO_4 0.07 (g/L), and $NaHCO_3$ 2.00 (g/L), pH 8.0] supplemented with yeast extract (0.04% w/v) at 22°C. For siderophores production, bacterial strains were cultivated at 22°C in GASN medium [2 g/L L-asparagine, 7 g/L glucose, 0.96 g/L Na_2HPO_4, 0.44 g/L KH_2PO_4, and 0.2 g/L $MgSO_4·7H_2O$, pH 7.0] [16].

Copper Ore Concentrate and Middlings

Copper ore concentrate and middlings (intermediate during ore processing) were received from KGHM S.A. "Polska Miedź" (Poland). Copper ore concentrate contained 185970 mg Cu/kg, 22169 mg Fe/kg, 2460 mg Pb/kg, 3765 mg As/kg, 533, 1 mg Ni/kg, and 1273 mg Co/kg. Middlings contained 21495 mg Cu/kg, 4290 mg Fe/kg, 291, 5 mg As/kg, 356, 5 mg Ni/kg, and 594 mg Co/kg. The main arsenic minerals are enargite (AsV), tennantite (AsIII), and realgar (AsIII). Arsenopyrite is present as trace mineral. Sandstone, dolomite, and limestone are gangue. Sulphur content is 0.3–0.9% (mainly as sulfides and sulfates).

Arsenopyrite

Crystals of arsenopyrite were received from ancient gold mine in Zloty Stok.

Arsenic Respiration Screening Test

The ability of bacterial isolates to reduce As (V) in respiratory processes was tested using MSM agar plate containing 5 mM sodium lactate as the carbon source and 5 mM sodium arsenate as the electron acceptor. Cultures were grown under anaerobic conditions and, after 7 days of cultivation at 22°C, the agar plates were flooded with 0.1 M $AgNO_3$ solution. The reaction between $AgNO_3$ and As (III) or As (V) results in the formation of a coloured precipitate [17]. A brownish precipitate reveals the presence of Ag_3AsO_4 (silver arsenate) in the medium, while a yellow precipitate shows the presence of (silver arsenite) (colonies expressing arsenate reductase).

Chemical Analyses

Solid samples (bacterial biomass, powders of copper ore concentrate, and middlings) were thoroughly dried at 60°C, and then 9 mL 65%HNO_3 and 1 mL 36%H_2O_2 were added to 0.25–0.30 g of the dry mass and digested in a closed system with heating in a microwave oven (Milestone Ethos Plus with Lab Terminal 800 Controller, Italy) [18]. Liquid samples (culture supernatants and siderophores solutions) were placed in 12 mL glass vials (Agillent, USA) and mixed with 65% HNO_3 in a ratio 4 : 1. Quantitative analysis of As, Cu, and Fe was performed by flame atomic absorption spectrometry (FAAS) and graphite furnace atomic absorption spectrometry (GFAAS) (AA Solaar M6 Spectrometer, TJA Solutions, UK) using standard solution (Merck, Darmstadt, Germany) prepared in 0.5 M HNO_3. Arsenic speciation in culture supernatants was determined as described by Drewniak et al. (2008) [19].

PCR Amplification and Sequencing: 16S rRNA Gene, arrA and arsC Genes

Amplification of the 16S rRNA genes and the arsC genes was performed as described by Drewniak et al. (2008) [20]. For the amplification of the arrA genes, primer pair ArrAfwd and ArrArev were used as described by Malasarn et al. (2004) [21]. The PCR products were ligated with the vector pGEM-T-Easy (Promega) and were transformed to chemically competent Escherichia coli TG1 cells. Plasmid inserts were sequenced on an ABI3730 DNA analyser (Applied Biosystems) at the Laboratory of DNA Sequencing and Oligonucleotide Synthesis, IBB PAS, using universal M13F and M13R primers. Additional primers 518F and 519R [22] were used for sequencing of the 16S rRNA genes. Partial sequences were assembled using Clone Manager Professional Suite software (version 8) and were verified manually.

Phylogenetic Analysis

The near full-length 16S rRNA gene sequences (~1.4 kbp) from the isolates and reference sequences from the known arsenic-utilizing bacteria were aligned using ClustalW software (http://www.ebi. ac.uk/clustalw). The obtained alignment was adjusted manually and then used to construct a phylogenetic tree. The unrooted tree was constructed using the distance matrix "Neighbor-Joining Method" with NEIGHBOR from the PHYLIP 3.6 software package [23]. Distance matrices were calculated by DNADIST (PHYLIP) using Jukes-Cantor formula. Bootstrap analysis was carried out 1000 times using SEQBOOT (PHYLIP). A consensus tree was computed with CONSENS (PHYLIP).

Determination of the Minimum Inhibitory Concentrations of Metals/Metalloids

To examine the minimal inhibitory concentrations (MIC) of heavy metals, 96-well microplates containing LB medium amended with the respective heavy metal compounds were used. Each well of the microplates was inoculated with cells from fresh overnight cultures to a final density of approximately 10^6 cells/mL and then incubated for 48 h at 22°C in aerobic conditions. The following metals and their compounds were used for MIC determination: As (III) 0.0–25.0 mM; As (V) 0.0–500 mM; Cu (II) 0.0–5.0 mM; Cd (II) 0.0–5.0 mM and Co (II) 0.0–5.0 mM; and Fe (III) 0.0–25.0 mM, Ni (II) 0.0–5.0 mM, Pb (II) 0.0–20 mM, and Zn (II) 0.0–5.0 mM. The MIC was defined as the lowest concentration of Me^{n+} that completely inhibited bacterial growth.

Detection of Siderophores and Their Chemical Nature

The production of siderophores was examined by aerobic growth experiment on GASN medium [16]. The concentration of siderophores present in culture supernatants was measured using the method described by Schwyn and Neilands [24]. A standard curve was prepared with deferoxamine mesylate (DFOB). The quantity of siderophores produced was determined from the standard curve using CAS assay solution [24] and the absorbance value measured at 630 nm after 1 h of incubation [25] and denoted as mM DFOB. Detection of hydroxamate siderophores was performed using $FeCl_3$ test [26] and for catecholates the Arnow test was used [27].

Test for Dissolution of Arsenic Minerals by Siderophores Produced by the Isolates

Modified method described by Drewniak et al. (2010) [10] was used for determination of bacterial siderophores ability to dissolve arsenopyrite (FeAsS) and arsenic minerals (enargite (Cu_3AsS_4) and tennantite (($Cu,Ag,Fe,Zn)_{12}As_4S$)) contained in copper ore concentrates and middlings. The following modifications were made: in dissolution reactions, 50 mL of 0.2 mM bacterial siderophores and sterile GASN medium (as a control) were added to 0.5 g of a given, sterile mineral, and the concentrations of As and Fe in the solution were measured after 48 h of incubation in aerobic conditions at 22°C using graphite furnace atomic absorption spectrometry (GFAAS) (AA Solaar M6 Spectrometer, TJA Solutions, UK). The method of siderophores preparation was as follows: (i) strains were grown in GASN medium for 24 h at room temperature in aerobic conditions, (ii) the cultures were then centrifuged (10,000 ×g, 10 min.), and the supernatants were used as siderophore preparations. Siderophores concentration was extrapolated from a standard curve and represented in mM of desferrioxamine B (DFOB).

Arsenic Minerals Utilization in Respiratory Processes

The ability of the isolates to utilize arsenic containing minerals (copper ore concentrates and middlings obtained from KGHM S.A, Poland) in respiratory processes was examined in the MSM medium supplemented with 5 mM sodium lactate as the electron donor and as the sole carbon source. Sterilized, powdered samples of minerals were added to the MSM medium (100 mL) to a final concentration of 1% (w/v) and were inoculated with cells from fresh overnight cultures to a final density of approximately 10^6 cells/mL. The cultures were incubated at 22°C for 21 days under anaerobic conditions, when the ability to use the investigated minerals as a final electron acceptor was tested. Noninoculated samples were used as controls. The cultures were sampled for chemical analyses (pH, heavy metal concentration, and arsenic speciation) and estimation of colony forming unit (CFU) at the start of the experiment and every 7 days.

Nucleotide Sequence Accession Number

The 16S rRNA gene sequence of isolated strains has been deposited in GenBank under accession numbers KF986639, KF986640, KF986641, and KF986642.

RESULTS AND DISCUSSION

Isolation and Identification of Dissimilatory Arsenate Reducers

Microbial mats form sediments sampled from the ancient Zloty Stok gold mine (SW Poland) were used as the source of arsenic-metabolizing microbes, for several reasons. Our previous studies [20] showed that the physical and chemical conditions prevailing in the Zloty Stok mine promote the growth and development of arsenic-transforming microbes. Arsenic in the mine is present in soluble form as arsenite and arsenate in mine waters and also occurs as primary (arsenopyrite and lollingite) and secondary minerals (such as scorodite) deposited in the rocks and sediments [28]. The mine waters and sediments also contain high amounts of other heavy metals (e.g., Cu, Co, Mn, and Zn) and therefore the microbial mats seemed to be an ideal source for the isolation of model dissimilatory arsenate reducing bacteria (DARB) that can be capable of dissolution of arsenic-bearing minerals. Regardless of the mechanism of arsenic release from polymetallic ores/minerals, microorganisms that are involved in dissolution processes should be adequately adapted to the surrounding conditions, for example, resistance to arsenic and heavy metals.

Microbial mats were inoculated into the modified MSM medium supplemented with 2.5 mM sodium arsenate and 5 mM sodium lactate. After 7 days of incubation under anaerobic conditions, enrichments were subcultured twice for 7 days each and the cultures were plated on MSM agar. Morphological observations (colony colours and cell shape), physiological test ($AgNO_3$ test), and 16S rRNA gene analysis (ARDRA analysis and sequencing) allowed for the identification of four different strains (OM1, OM2, OM4, and OM17) capable of anaerobic growth with As (V) reduction. BLASTN analysis of partial 16S rDNA sequences (~1.4 kbp) showed that all the isolates belonged to the class γ-Proteobacteria. The nearest known phylogenetic relative of the strain OM1, with a sequence similarity of 99%, is Shewanella sp. OTUC2. The strain OM2 showed the highest similarity (99%) to genus Pseudomonas koreensis strain JH18. The strain OM4 was closely related (100% of similarity) to genus Aeromonas hydrophila, whereas strain OM17 was closely related (99% of similarity) to Serratia liquefaciens ATCC 27592. Phylogenetic studies confirmed BLASTN analysis and showed that 16S rRNA gene sequences of all isolates are located in clusters of closely related species within the class γ-Proteobacteria (Figure 1). Interestingly, two strains, Aeromonas sp. OM4 and Serratia sp. OM17, are the first described representatives of their genera, which are capable of dissimilatory arsenate reduction. So far described Aeromonas spp. and Serratia spp. strains were able to reduce arsenate, but only using cytoplasmic reductase as a resistance mechanism [20, 29, 30].

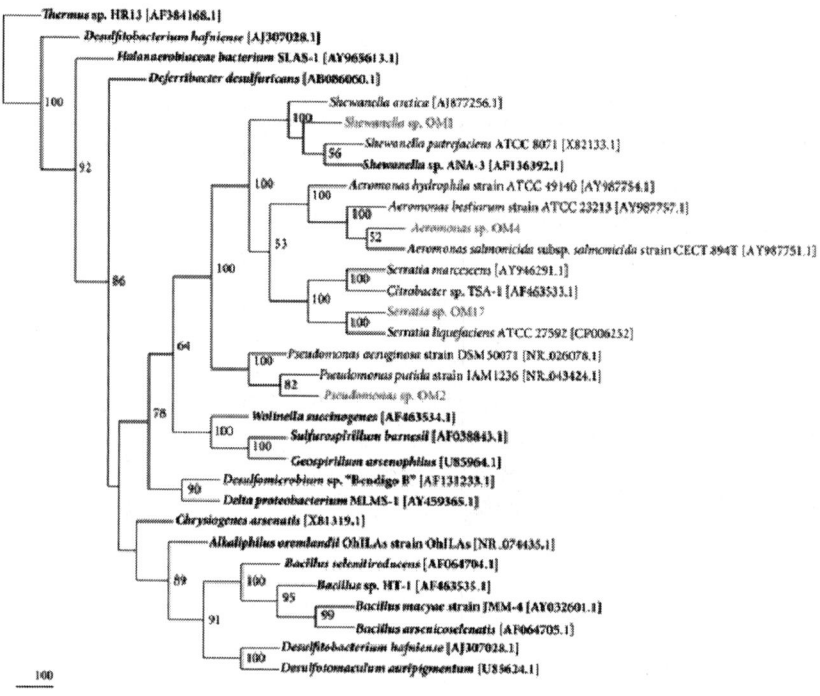

Figure 1: Neighbour-joining tree, computed based on the alignment of 16S rRNA gene sequences, showing the phylogenetic relationship of arsenic-respiring isolates with known arsenic resistant and dissimilatory arsenate reducing bacteria. The analysis included the data from ~1400 nucleotide positions. The tree is a consensus of 100 neighbor-joining trees. Percentage values on each branch represent the corresponding bootstrap probability values obtained in 100 replications and numbers are shown only for values of >50%. Sequences derived from arsenic-respiring isolates (OM1, OM2, OM4, and OM17) are indicated in large bold red type; sequences derived from other

dissimilatory arsenate reducing bacteria are indicated in black bold type. The remaining sequences indicated in normal type derived from strains that are only resistant to arsenic.

All of the tested bacteria showed a broad range of temperature tolerance: 4–37°C, with the optimum at 30°C (OM1, OM2, and OM17) and 37°C (OM4) (Table 1). The pH optimum was found to be slightly acidic or close to neutral (4–7), but growth at alkaline pH (10) was also observed (Table 1). Ability to grow in a broad range of pH and temperature conditions makes the isolated strains potential biotechnological tools for bioremediation of arsenic-contaminated sites. One of the main factors limiting the use of bacteria in bioremediation technologies is a sensitivity of strains to changing physicochemical conditions. The broad spectrum of tolerance to pH and temperature increases the chances of survival in the environment.

Table 1: Tolerance to arsenic and arsenate reductase activity and the presence of arsenic metabolism genes

Strain	Temperature (°C)		pH		MIC for AsV (mM)	MIC for AsIII (mM)	Presence of genes coding for cytoplasmic and respiratory arsenate reductases	
	Range	Optimum	Range	Optimum			arsC	arrA
Shewanella sp. OM1	4–37	30	5–10	7	350	10	+	+
Pseudomonas sp. OM2	4–37	30	5–9	5	250	14	+	+
Aeromonas sp. OM4	4–37	37	5–8	5	350	16	+	+
Serratia sp. OM17	4–37	30	4–10	4	400	12	+	+

Detoxification Mechanisms

The primary role of microorganisms living under unfavourable environmental conditions is to survive. Detoxification mechanisms that protect against harmful substances are very often connected with processes that require energy supply and sometimes are associated with respiratory processes in which energy is generated. The most common mechanism of detoxification is based on blocking the membrane channels through which toxic substances enter the cell. Another method of protection relies on the active removal out of the cell using specific membrane pumps [31]. Arsenite ions are directly removed by membrane permeases, while arsenates ions require transformation into the arsenites prior to removal. In this study we have verified the level of arsenic resistance and we investigated the presence of cytoplasmic arsenate reductase genes (arsC) in the genomes of the isolated strains.

The isolates were hypertolerant to arsenate (250–400 mM) and showed high resistance to arsenite (10–16 mM).Serratia sp. OM17 was the most resistant to As (V) isolate, which tolerated concentrations of arsenate up to 400 mM. The highest MIC for As (III) was found for Aeromonas sp. OM4, which tolerated concentrations of arsenite up to 16 mM. Such high resistance to arsenic was also found in other arsenic-tolerating microorganisms, but any described arsenic hypertolerant strain was able to use As (V) in dissimilatory arsenate reduction [20, 32]. The presence of arsC gene was confirmed in the genomes of all of the tested strains (Table 1). BLASTN results showed that arsC genes of Shewanella sp. OM1, Pseudomonas sp. OM2, and Serratia sp. OM17 have a high similarity to cytoplasmic arsenate reductases genes of closely related species. Nucleotide sequence of arsC gene of Shewanella sp. OM1 is 78% identical to the arsenate reductase gene of Shewanella putrefaciens CN-32. The arsC sequence of Pseudomonas sp. OM2 is 84% identical to the arsenate reductase gene of Pseudomonas denitrificans ATCC 13867, while the arsC of Serratia sp. OM17 is 96% identical to the arsenate reductase gene of Serratia liquefaciens ATCC 27592. In turn, the arsC gene ofAeromonas sp. OM4 is 76% identical to arsenate reductase of Acidovorax ebreus TPSY. Phylogenetic analysis showed that the arsC genes of the OM isolates are closely related to the arsenate reductase gene sequences of known arsenic resistant strains, but not necessarily from the same species (see Figure S1 available online athttp://dx.doi.org/10.1155/2014/841892).

In the context of arsenic minerals dissolution, equally important as arsenic resistance is the tolerance to other heavy metals, such as Cu, Co, Cd, Zn, Fe, Ni, and Pb. The level of resistance to the tested metals was between 0.25 and 4 mM: Cu (II) (for all strains up to 4 mM), Co (II) (OM1 and OM17 0.75 mM, OM2 0.5 mM, and OM4 1 mM), Cd (II) (OM1, OM2, and OM4 0.25 mM and OM17 0.5 mM), Zn (II) (OM1 and OM 17 up to 3.5 mM, OM2 1 mM, and OM4 0.75 mM), Fe (III) (for all the strains up to 2 mM), Ni (III) (OM1 and OM4 up to 3 mM and OM2 and OM17 up to 2.5 mM), and Pb (II) (for all the strains up to 2 mM). Such heavy metal multiresistance enables their active growth in environments rich in heavy metals.

Description of the Process That Promotes Arsenic Mobilization

The mechanisms of microbial arsenic mobilization depend on many biogeochemical factors, among which the most important seem to be the following: the presence of the appropriate physiological group of microbes, type of minerals (primary or secondary), pH (acidic or neutral), Eh (reductive or oxidative conditions), and the availability of oxygen or other electron acceptors. Direct mechanisms of arsenic mobilization include oxidative dissolution of primary arsenic minerals and reductive dissolution of secondary arsenic minerals [6].

Arsenate Respiration Process

To test the isolates for the ability to grow by arsenate respiration, anaerobic growth experiments were conducted. All of the isolates grew with 5 mM sodium lactate as the electron donor and 2.5 mM sodium arsenate as the electron acceptor (Figure 2). In almost all of the cultures (except Aeromonas sp. OM4) arsenate reduction was observed during the exponential phase, in which the growth rate and the rate of As (V) reduction were proportional. Growth did not occur when medium without arsenate was used or arsenate was completely utilized, indicating that arsenate was required for growth. The Serratia sp. OM17 strain was found to be the most efficient isolate, in terms of the highest arsenate reduction rate [7.81 mg As (V) $\cdot L^{-1} \cdot h^{-1}$]. Complete reduction of 2.5 mM (187.5 mg$\cdot L^{-1}$) arsenate into arsenite was observed after 48 h of incubation. Other strains had a slightly lower arsenate reduction rate, and the complete transformation of As (V) into As (III) required at least 72 hours of incubation. Shewanella sp. OM1 and Pseudomonas sp. OM2 completely reduced 2.5 mM of As (V) to As (III) within 72 hours, while Aeromonas sp. OM4 within 96 hours. In addition to the lowest As (V) reduction rate, Aeromonas sp. OM4 was not able to completely reduce arsenates.

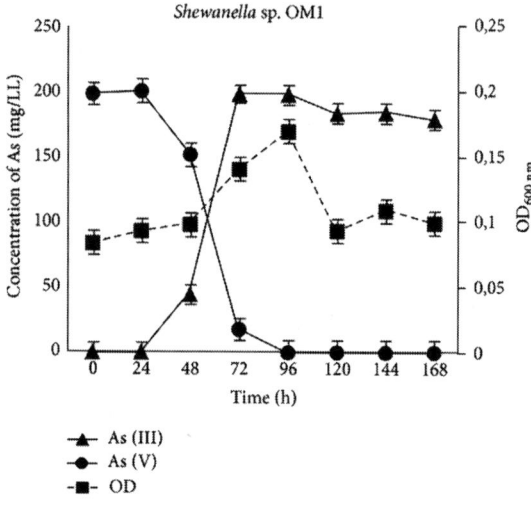

(a)

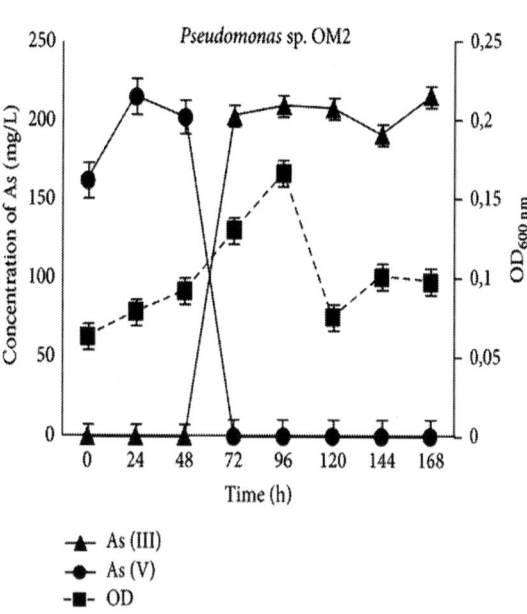

(b)

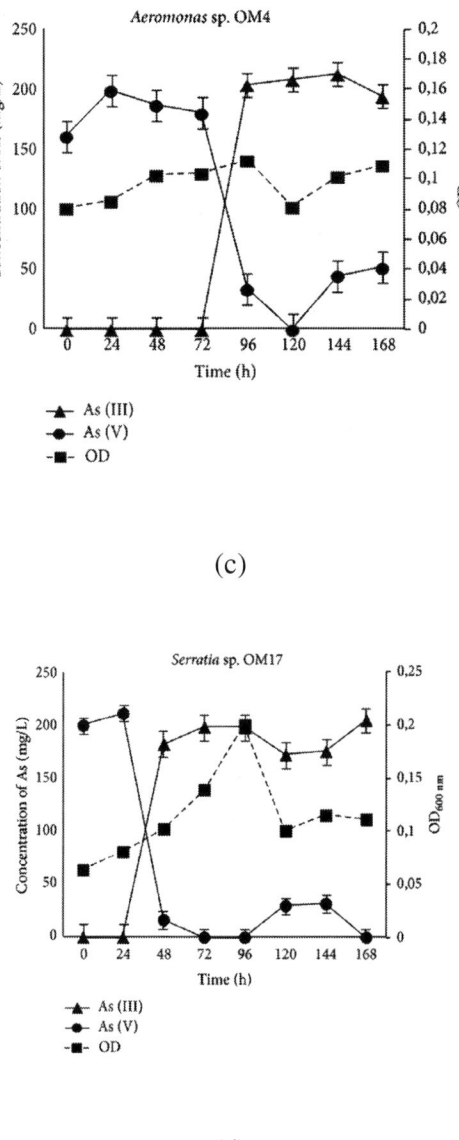

(c)

(d)

Figure 2: Dissimilatory arsenate reduction mediated by Shewanella sp. OM1, Pseudomonas sp. OM2, Aeromonas sp. OM4, and Serratia sp. OM17. Arsenate respiratory processes were tested using cultures grown under anaerobic conditions in the MSM medium, supplemented with 2.5 mM sodium arsenate and 5 mM sodium lactate.

All of the tested strains were unable to reduce the arsenate at concentrations above 2.5 mM, despite their ability to tolerate high concentrations of As (III) (10–16 mM) and As (V) (250–400 mM). Interestingly, most of dissimilatory arsenate reducers isolated during similar studies were able to reduce arsenate at much higher concentrations, exceeding 10 mM [11, 33] and sometimes reaching up to 40 mM [34], although they come from environments in which the arsenic is present at a similar level as in Zloty Stok gold mine (isolation site of the tested strains).

Similar to other known arsenic respiratory strains, the arrA gene (coding for the large subunit of dissimilatory arsenate reductase) was found in the genomes of all of the isolates. BLASTN results showed that all dissimilatory arsenate reductase genes of the isolates (OM1, OM2, OM4, and OM17) are closely related to arrAgenes detected in metagenomic studies and described as uncultured bacterium clones. In turn, phylogenetic studies with the use of arrA sequences of the isolates and known arsenic respiratory bacteria showed that arsenic respiratory genes of the OM (1, 2, 4, and 17) isolates form a common, separate cluster and they are not closely related to the genes of their phylogenetic relatives.

Dissolution of Arsenic Containing Minerals by Metabolites Produced by DARB

Nutrient uptake from minerals may be performed by different mechanisms. Microorganisms may cause disaggregation of minerals through colonization and physical penetration of the mineral surface [35]. Extraction of vital elements from the crystal lattice of minerals may be also supported by organic agents produced by the cells. Among the most common microbial organic agents involved in minerals dissolution are siderophores: high-affinity, metal-binding compounds secreted outside the cell envelope that can chelate metal ions and bind to atoms on the mineral surface. They can effectively bind many metal cations including Fe, Mg, Mn, Cr, Ga, Pl, Pb, Cd, Zn, Cu, Ni, U, Co, Sn, and As [36–40].

Biochemical tests showed that under iron-limiting conditions all the DARB isolates produce siderophores that are classified to the hydroxamate-type. The highest concentration of siderophores in GASN medium after 48 h of incubation at 22°C was noted in the culture of Aeromonas sp. OM4 (118 µM) and the lowest in theShewanella sp. OM1 culture (88 µM). Pseudomonas sp. OM2 and Serratia sp. OM17 produced siderophores at similar levels, 116 µM and 110 µM, respectively.

A further issue that needs clarification is the ability of the siderophores, produced by the selected strains, to dissolve (i) arsenopyrite and (ii) middlings and copper concentrates containing arsenic minerals (enargite, realgar, and tennantite). For this purpose, cell-free metabolites (containing siderophores) obtained from the cultivation of the DARB isolates under iron and arsenic limiting conditions in GASN medium were used.

After 48 h of leaching of arsenic minerals by the DARB siderophores, iron and arsenic were released into the solutions of all of the tested samples. Siderophores produced by most of the tested strains show the capacity to release arsenic from minerals at a level similar to the control. In the case of arsenopyrite dissolution by siderophores, only metabolites produced by Serratia sp. OM17 showed significant differences in the release of arsenic in relation to the control sample. The concentration of arsenic in the OM17 siderophores solution was almost eleven times higher than in the control sample (4.07 mg/L of As was noted in Serratia sp. OM17 siderophores solution, whereas 0.37 mg/L in sterile GASN medium). Twice as high concentration of arsenic than in the control was observed in the siderophores solution of the Aeromonas sp. OM4 strain. The concentration of arsenic mobilized from arsenopyrite by siderophores produced by two other strains (OM1 and OM2) was slightly higher (1.37 and 1.8 times, resp.) than in the control (Figure 3(a)). The release of arsenic from copper concentrate and middlings by siderophores was inefficient and arsenic concentrations were at the level of the control. The highest concentration of arsenic in the middlings dissolution was observed when the siderophores solution of Shewanella sp. OM1 was used (1.47 times higher than the control) (Figure 3(b)). In the case of copper ore concentrates, the highest concentration of arsenic was observed when siderophores solution of Serratia sp. OM17 was used (1.66 times higher than in the control) (Figure 3(c)).

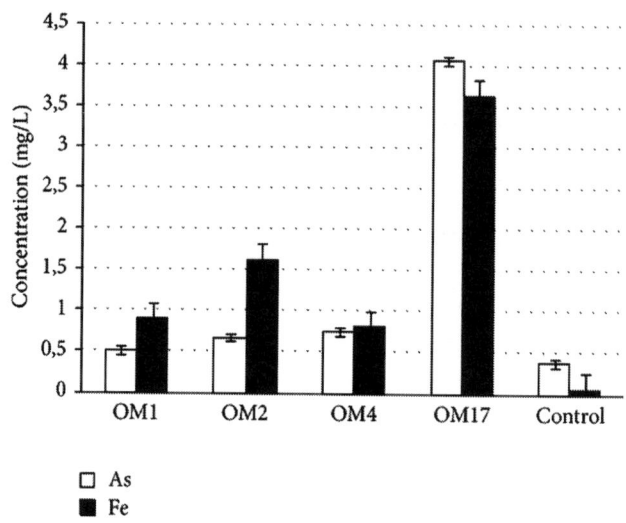

(a)

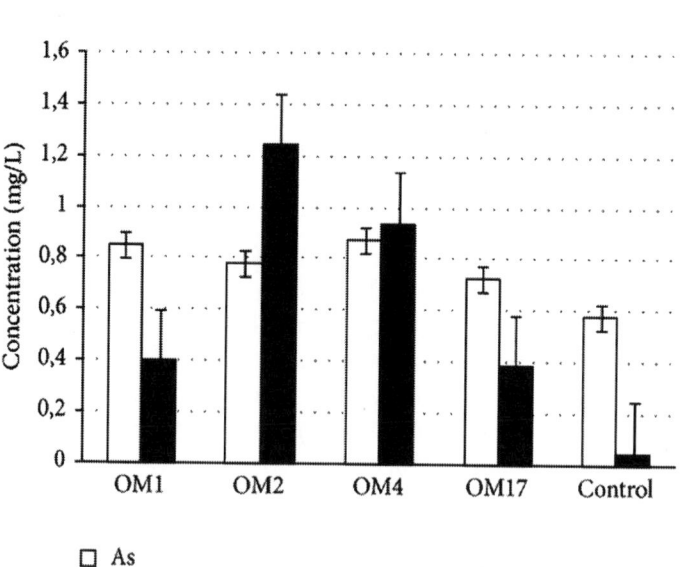

(b)

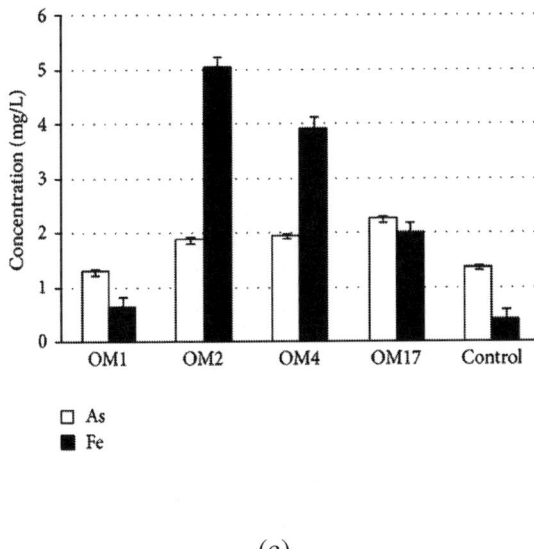

(c)

Figure 3: Release of arsenic and iron from arsenopyrite (a), middlings (b), and copper ore concentrates (c), after 48 h of incubation with sterile GASN medium as a control, and siderophores produced by the isolated strains: Shewanella sp. OM1, Pseudomonas sp. OM2, Aeromonas sp. OM4, and Serratia sp. OM17.

In contrast to arsenic case, the iron concentration in almost all of the samples was at a much higher level than in the control. The highest iron concentration in arsenopyrite sample was observed in the siderophores solution of Serratia sp. OM17 (Figure 3(a)), in which the concentration of Fe exceeded ~72 times the iron content recorded in the control sample. In the middlings and copper ore concentrates samples, the highest iron concentrations were observed when the siderophores solution of Pseudomonas sp. OM2 was used (25 times and 12.5 higher than the control in the middlings and copper concentrates, resp.) (Figures 3(b) and3(c)). Regardless of the copper mineral, the lowest concentration of iron was noted in the case of siderophores produced by Shewanella sp. OM1 (only 1.5 times more than in the control with copper concentrates) (Figure3(c)).

The obtained results demonstrate that siderophores produced by dissimilatory arsenate reducers are involved in dissolution of arsenic containing minerals. Based on the results it can be concluded that the mechanism of arsenic minerals dissolution can be driven by iron uptake through siderophores, while arsenic is released simultaneously as by-product. These results are consistent with the previously described arsenic mobilization mechanisms used by dissimilatory iron reducers [14, 41].

Removal of Arsenic from As-containing Minerals

It is commonly known that the release of arsenic from As-bearing minerals may take place by two main processes: (i) oxidation of the primary arsenic minerals and (ii) reductive dissolution of the secondary minerals. Primary As-minerals such as arsenopyrite or realgar can be dissolved by direct or indirect mechanisms in the presence of oxygen by chemolithoautotrophic microorganisms [42]. In turn, under reducing conditions, microorganisms can use arsenate adsorbed on the surface of iron minerals (e.g., scorodite or ferrihydrite) as a terminal electron acceptor in arsenic respiration [14, 43].

In the literature, there is not much information concerning the release of arsenic from the primary minerals under reducing conditions. Thus, it is interesting to investigate if under anaerobic conditions bacteria are capable of arsenic mobilization from primary minerals and in particular whether they are able to use such minerals in the process of respiration as an electron donor. In order to verify the above hypothesis we performed respiration experiments using primary arsenic minerals as the final electron acceptors. Arsenopyrite (FeAsS), as one of the most common primary arsenic minerals, containing arsenic in reduced form, was used in this experiment as a control in which bacterial cells have not final electron acceptor. Copper concentrates and middlings were used as a source of primary arsenic mineral (such as enargite), in which arsenic occurs as arsenate. Both Cu-sources are polymetallic and contain arsenic as impurities (0.37% for the concentrates and 0.029% for the middlings). The ability of the isolates to utilize arsenic containing minerals in respiratory processes was examined in the MSM medium supplemented with 5 mM lactate as the electron donor and the sole carbon source and 1% of appropriate mineral as a source of electron donors.

As predicted, none of the isolated strains were capable of using arsenopyrite as a respiratory substrate. Growth was not observed in any culture, but only death phase. Number of colony forming units decreased from the initial 10^6 cfu/mL to 10^2–10^3 cfu/mL after 21 days of incubation in all cultures. There were no significant changes in pH. After 7 days of incubation pH in all cultures increased to ~8.5 and remained stable until the end of the experiment (in control sample the pH was stable throughout the experiment). The concentrations of arsenic and iron in the supernatants exceeded the levels observed in the control sample (sterile medium), but the amount of mobilized arsenic and iron was low (Figure 4), especially if the total arsenic and iron content in arsenopyrite was taken into account. The mobilization of small amounts of As and Fe from FeAsS by the cells being in decline phase can be explained by an unspecific dissolution of arsenopyrite by metabolites (e.g., ligands and organic acids) released during cell lysis. Thus dissolution in reductive conditions mediated by dissimilatory arsenate reducing bacteria is not the main driving force leading to arsenic mobilization from arsenopyrite, but only a passive process that should be taken into account during long-term environmental risk assessment. These results are in accordance with the current knowledge about the microbiological dissolution of the primary arsenic minerals; that is, the primary arsenic minerals are mainly utilized under aerobic conditions by oxidative dissolution [6].

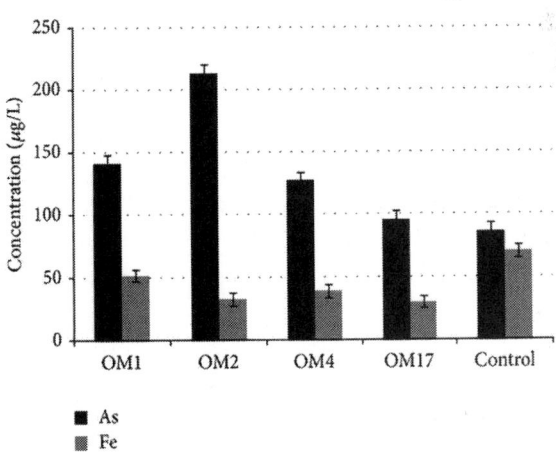

Figure 4: Arsenopyrite dissolution under anaerobic conditions by dissimilatory arsenate reducing bacteria: Shewanella sp. OM1, Pseudomonas sp. OM2,

Aeromonas sp. OM4, andSerratia sp. OM17. As and Fe concentrations in culture liquid after 21 days of incubation under anaerobic condition in MSM medium supplemented with 1% FeAsS and 5 mM sodium lactate.

A quite different effect was observed in the respiration experiments with arsenic containing copper sulfide minerals used as the final electron acceptor. All of the strains were capable of growing under anaerobic conditions in the MSM medium containing (1%) sterile Cu-minerals as the sole electron acceptor and considerable arsenic concentrations were detected in all of the culture supernatants after 21 days of incubation (Figure 5). In almost all of the cultures, the concentration of arsenic in the culture supernatants increased in time (Figures 5(a1) and 5(b1)) and correlated with the rates of growth (data not shown). The final amount of the removed arsenic oscillated between 19.28 and 81.95 mg/kg (6.6% and 28.1%) for the middlings (Figure5 (a1)) and 7.93 and 76.75 mg/kg (0.21% and 2.03%) for copper ore concentrates (Figure 5(b1)). The most efficient strain, in terms of arsenic release, was Aeromonas sp. OM4 strain cultured in middlings and copper ore concentrate (As concentration in culture liquid reached 819.50 and 922.50 µg/L, resp.). The lowest level of arsenic removal was observed for Shewanella sp. OM1 culture (As concentration in culture liquid was 192.75 µg/L for middlings and 678.75 µg/L for copper ore concentrates) (Figures 5(a1) and 5(b1)). Irrespective of which strain was used, the effectiveness of arsenic extraction was 10 times higher in the middlings than in the copper ore concentrates leaching experiment. The maximum recovery of arsenic from middlings was 28.11% after 21 days for Aeromonas sp. OM4 (Figure 6), while the maximum recovery of arsenic from copper ore concentrates was 2.47% at 21 days for Pseudomonas sp. OM2 (Figure 6). For all of the strains, the arsenic recovery process was selective and was not associated with simultaneous removal of copper (Figures 5(a2) and 5(b2)). High concentration of copper was noted in all of the samples at the beginning of cultivation (Figures 5(a2) and 5(b2)), which was probably associated with chemical leaching of copper (by medium components and residual metabolites from the overnight cultures). Moreover, in all of the cultures and control samples, copper concentration decreased with time. Only in the case of Serratia sp. OM17 cultured on middlings, elevated level of copper leaching (767.50 µg/kg) was reported after 21 days of incubation. However, it should be noted that the resulting value represents only 0.04% of the initial copper content. These results may indicate the slow copper precipitation process under neutral conditions (pH fluctuated around 7.5 ± 0.4 during the experiment) and absence of microbial activity connected with the leaching of copper.

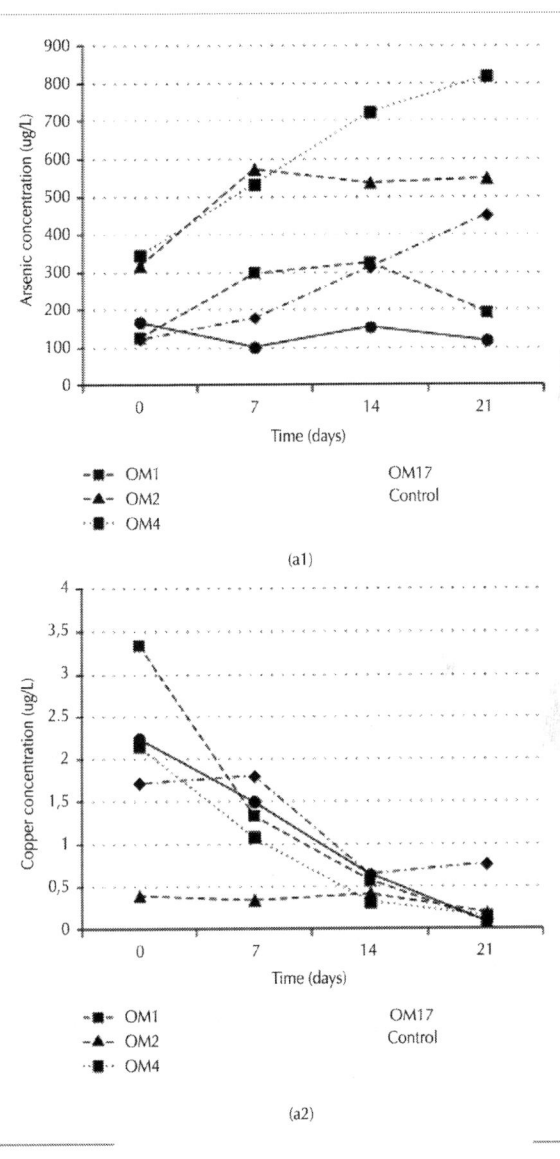

(a1)

(a2)

(a)

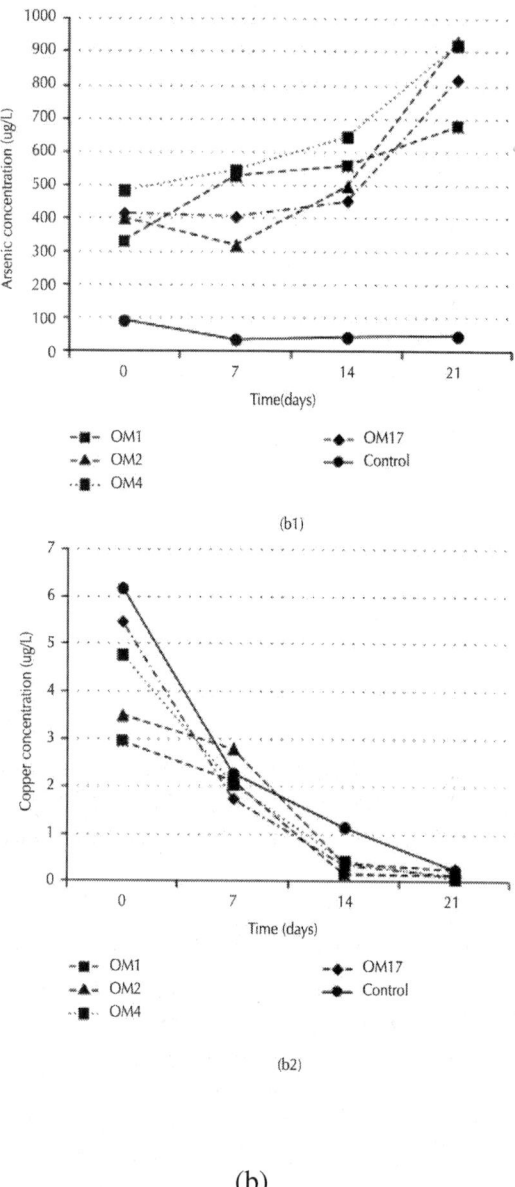

(b)

Figure 5: Dissolution of As-bearing copper minerals mediated by dissimilatory arsenate reducing bacteria. Arsenic (1) and copper (2) release during 21 days of incubation in MSM medium supplemented with 5 mM sodium lactate and (a) 1% middlings and (b) 1% copper ore concentrates.

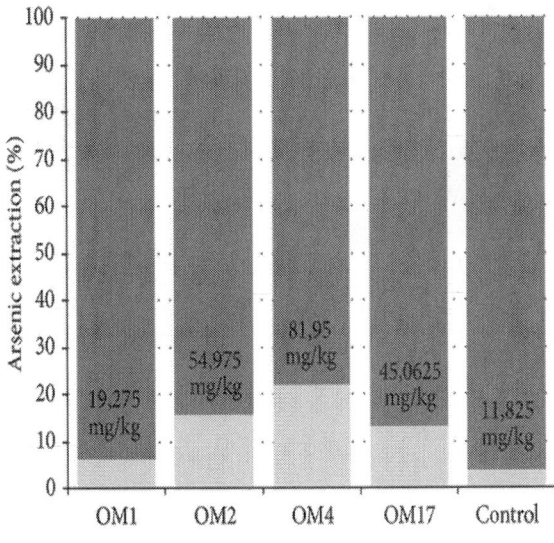

(a)

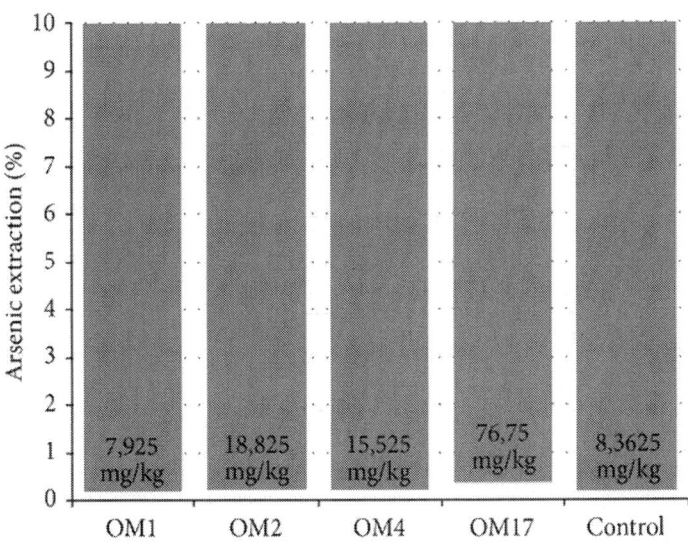

(b)

Figure 6: Comparison of effectiveness of arsenic removal from middlings (a) and copper ore concentrates (b) mediated by dissimilatory arsenate reducing bacteria: Shewanella sp. OM1, Pseudomonas sp. OM2, Aeromonas sp. OM4, and Serratia sp. OM17.

On the basis of the above results we can postulate that arsenic can be released from some As-bearing polymetallic minerals (such as copper ore concentrates or middlings) in reductive conditions by dissimilatory arsenate reducers. However, the mechanisms of arsenic release are still unknown. Comparing the data from the arsenopyrite and copper concentrates/middlings respiration experiment, we can only assume that the mechanism of arsenic release from Cu-minerals is indirect. To understand how the DARPs release arsenic from minerals, deep physiological and microbe-mineral interaction (including arsenic speciation) analysis is needed. The ability to use different inorganic electron donors (e.g., Fe^{2+} or S^{2-}), which may be conjugated to electron acceptors other than oxygen (e.g. NO_3^-, and $As_3O_5^{3-}$), requires verification. Similarly, the hypothesis that bacteria can utilize polymetallic sulfide as a source of enzyme cofactors such as Mo, Cu, Zn, and Mg [35] must be also verified.

CONCLUSIONS

In this work, we have shown that dissimilatory arsenate reducing bacteria have a great potential for arsenic mobilization into the environment, not only from secondary arsenic minerals but also from the primary arsenic minerals deposited in polymetallic ores. All of the isolated strains give positive results in the arsenic respiratory test. They effectively grew with sodium lactate as the electron donor and sodium arsenate as the electron acceptor. Resistance to high concentrations of arsenite and arsenate, as well as a number of other heavy metal compounds, was another common feature of the isolates. Moreover, all of the strains displayed the ability to grow in a broad range of pH values and temperatures and under both aerobic and anaerobic conditions. Under iron-limiting conditions all of the isolated DARBs produced siderophores that were involved in dissolution of arsenic containing minerals. None of the isolated strains were capable of using arsenopyrite as a respiratory substrate, but (low efficient) arsenopyrite dissolution in reductive conditions was observed. We postulate that it was only as a passive process, in which metabolites (e.g., ligands and organic acids) released during cell lysis may play a key role. In turn, utilization of As-bearing copper minerals in respiratory processes was efficiently carried out by most of the isolates. The DARBs strains were capable of growing under anaerobic conditions using As-bearing polymetallic minerals as the sole electron acceptor. Arsenic release from Cu-minerals was selective (not associated with simultaneous removal of copper) and as we concluded the process was indirect. This work clearly showed that the release of arsenic from primary As-minerals included in the polymetallic ores or wastes from their processing (such as copper concentrates and middlings) is possible under reducing conditions, but the mechanism is still unknown and requires further, detailed mineral-microbe interaction studies (to complete understanding).

ACKNOWLEDGMENTS

This work was supported by the Polish Ministry of Science and Higher Education in the form of Grant Juventus Plus no. 0079/P01/201070.

REFERENCES

1. D. J. Vaughan, "Arsenic," Elements, vol. 2, no. 2, pp. 71–75, 2006.

2. T. H. Hoang, J.-Y. Kim, and S. Bangb, "Source and fate of As in the environment," Geosystem Engineering, vol. 13, no. 1, pp. 35–42, 2010. ·

3. N. L. Piret, "Removal and safe disposal of arsenic in copper processing," JOM, vol. 51, no. 9, pp. 16–17, 1999.

4. D. Filippou, P. St-Germain, and T. Grammatikopoulos, "Recovery of metal values from copper—arsenic minerals and other related resources," Mineral Processing and Extractive Metallurgy Review, vol. 28, no. 4, pp. 247–298, 2007. · ·

5. J. P. Demaerel, "The behaviour of arsenic in the electrorefining process," in The Electrorefining and Winning of Copper, H. J. E, Ed., pp. 195–209, TMS, Warrendale, Pa, USA, 1987.

6.   L. Drewniak and A. Sklodowska, "Arsenic-transforming microbes and their role in biomining processes," Environmental Science and Pollution Research International, vol. 20, no. 11, pp. 7728–7739, 2013. ·

7.   M.-N. Collinet and D. Morin, "Characterization of arsenopyrite oxidizing Thiobacillus. tolerance to arsenite, arsenate, ferrous and ferric iron," Antonie van Leeuwenhoek, International Journal of General and Molecular Microbiology, vol. 57, no. 4, pp. 237–244, 1990. · ·

8.   D. B. Johnson and K. B. Hallberg, "The microbiology of acidic mine waters," Research in Microbiology, vol. 154, no. 7, pp. 466–473, 2003. · ·

9.   E. D. Rhine, K. M. Onesios, M. E. Serfes, J. R. Reinfelder, and L. Y. Young, "Arsenic transformation and mobilization from minerals by the arsenite oxidizing strain WAO," Environmental Science and Technology, vol. 42, no. 5, pp. 1423–1429, 2008. · ·

10.   L. Drewniak, R. Matlakowska, B. Rewerski, and A. Sklodowska, "Arsenic release from gold mine rocks mediated by the activity of indigenous bacteria," Hydrometallurgy, vol. 104, no. 3-4, pp. 437–442, 2010. · ·

11.   A. M. Laverman, J. S. Blum, J. K. Schaefer, E. J. P. Phillips, D. R. Lovley, and R. S. Oremland, "Growth of strain SES-3 with arsenate and other diverse electron acceptors," Applied and Environmental Microbiology, vol. 61, no. 10, pp. 3556–3561, 1995.

12.   T. Krafft and J. M. Macy, "Purification and characterization of the respiratory arsenate reductase of Chrysiogenes arsenatis," European Journal of Biochemistry, vol. 255, no. 3, pp. 647–653, 1998.

13.   D. K. Newman, E. K. Kennedy, J. D. Coates et al., "Dissimilatory arsenate and sulfate reduction in Desulfotomaculum auripigmentum sp. nov," Archives of Microbiology, vol. 168, no. 5, pp. 380–388, 1997. · ·

14.   J. Zobrist, P. R. Dowdle, J. A. Davis, and R. S. Oremland, "Mobilization of arsenite by dissimilatory reduction of adsorbed arsenate," Environmental Science and Technology, vol. 34, no. 22, pp. 4747–4753, 2000. · ·

15. J. Sambrook and D. W. Russell, Molecular Cloning: A Laboratory Manual, Cold Spring Harbor Laboratory Press, New York, NY, USA, 2001.

16. A. Bultreys and I. Gheysen, "Production and comparison of peptide siderophores from strains of distantly related pathovars of Pseudomonas syringae and Pseudomonas viridiflava LMG 2352," Applied and Environmental Microbiology, vol. 66, no. 1, pp. 325–331, 2000.

17. D. D. Simeonova, D. Lièvremont, F. Lagarde, D. A. E. Muller, V. I. Groudeva, and M.-C. Lett, "Microplate screening assay for the detection of arsenite-oxidizing and arsenate-reducing bacteria,"FEMS Microbiology Letters, vol. 237, no. 2, pp. 249–253, 2004. · ·

18. J. Namieśnik, B. Zygmunt, and A. Jastrzębska, "Application of solid-phase microextraction for determination of organic vapours in gaseous matrices," Journal of Chromatography A, vol. 885, no. 1-2, pp. 405–418, 2000. · ·

19. L. Drewniak, R. Matlakowska, and A. Sklodowska, "Arsenite and arsenate metabolism of sinorhizobium sp. M14 living in the extreme environment of the zloty stok gold mine," Geomicrobiology Journal, vol. 25, no. 7-8, pp. 363–370, 2008. · ·

20. L. Drewniak, A. Styczek, M. Majder-Lopatka, and A. Sklodowska, "Bacteria, hypertolerant to arsenic in the rocks of an ancient gold mine, and their potential role in dissemination of arsenic pollution,"Environmental Pollution, vol. 156, no. 3, pp. 1069–1074, 2008. · ·

21. D. Malasarn, C. W. Saltikov, K. M. Campbell, J. M. Santini, J. G. Hering, and D. K. Newman, "arrA is a reliable marker for As(V) respiration," Science, vol. 306, no. 5695, p. 455, 2004. · ·

22. D. J. Lane, "16S/23S rRNA sequencing," in Nucleic Acid Techniques in Bacterial Systematics, E. Stackebrandt and M. Goodfellow, Eds., pp. 115–175, Wiley, New York, NY, USA, 1991.

23. J. Felsenstein, PHYLIP (Phylogeny Inference Package), Department of Genome Sciences, University of Washington, Seattle, Wash, USA, 2005.

24. B. Schwyn and J. B. Neilands, "Universal chemical assay for the detection and determination of siderophores," Analytical Biochemistry, vol. 160, no. 1, pp. 47–56, 1987.

25. F. A. Fekete, V. Chandhoke, and J. Jellison, "Iron-binding compounds produced by wood decaying basidiomycetes," Applied and Environmental Microbiology, vol. 55, no. 10, pp. 2720–2722, 1989.

26. J. B. Neilands, "Microbial iron compounds," Annual Review of Biochemistry, vol. 50, pp. 715–731, 1981.

27. L. E. Arnow, "Colorimetric determination of the components of 3,4-dihydroxyphenylalanine tyrosine mixtures," The Journal of Biological Chemistry, vol. 118, p. 531, 1937.

28. A. Chlebicki, B. Godzik, M. W. Lorenc, and A. Skłodowska, "Fungi and arsenic-tolerant bacteria in the hypogean environment of an ancient gold mine in Lower Silesia, SW Poland," Polish Botanical Studies, vol. 19, pp. 81–95, 2005.

29. C. R. Anderson and G. M. Cook, "Isolation and characterization of arsenate-reducing bacteria from arsenic-contaminated sites in New Zealand," Current Microbiology, vol. 48, no. 5, pp. 341–347, 2004. · ·

30. N. El Aafi, F. Brhadaa, M. Daryb, et al., "Rhizostabilization of metals in soils using Lupinus luteus inoculated with the metal resistant rhizobacterium Serratia sp. MSMC541," International Journal of Phytoremediation, vol. 14, no. 3, pp. 261–274, 2012. ·

31. S. Silver and L. T. Phung, "A bacterial view of the periodic table: genes and proteins for toxic inorganic ions," Journal of Industrial Microbiology and Biotechnology, vol. 32, no. 11-12, pp. 587–605, 2005. · ·

32. D. N. Joshi, S. J. S. Flora, and K. Kalia, "Bacillus sp. strain DJ-1, potent arsenic hypertolerant bacterium isolated from the industrial effluent of India," Journal of Hazardous Materials, vol. 166, no. 2-3, pp. 1500–1505, 2009. · ·

33. S. O. Soda, S. Yamamura, H. Zhou, M. Ike, and M. Fujita, "Reduction kinetics of As (V) to As (III) by a dissimilatory arsenate-reducing bacterium, Bacillus sp. SF-1," Biotechnology and Bioengineering, vol. 93, no. 4, pp. 812–815, 2006. · ·

34. D. Y. Sorokin and G. Muyzer, "Desulfurispira natronophila gen. nov. sp. nov.: An obligately anaerobic dissimilatory sulfur-reducing bacterium from soda lakes," Extremophiles, vol. 14, no. 4, pp. 349–355, 2010. · ·

35. K. Konhauser, Introduction To Geomicrobiology, Blackwell, Oxford, UK, 2007.

36. A. M. Aiken, B. M. Peyton, W. A. Apel, and J. N. Petersen, "Heavy metal-induced inhibition ofAspergillus niger nitrate reductase: applications for rapid contaminant detection in aqueous samples,"Analytica Chimica Acta, vol. 480, no. 1, pp. 131–142, 2003. · ·

37. A. Yarnell, "Nature›s tiniest geoengineers," Chemical and Engineering News, vol. 81, no. 42, pp. 24–25, 2003.

38. A. Malik, "Metal bioremediation through growing cells," Environment International, vol. 30, no. 2, pp. 261–278, 2004. · ·

39. B. Peyton and W. A. Apel, "Siderophore influence on the mobility of both radionuclides and heavy metals," INRA Informer, vol. 5, pp. 2–4, 2005.

40. A. Nair, A. A. Juwarkar, and S. K. Singh, "Production and characterization of siderophores and its application in arsenic removal from contaminated soil," Water, Air, and Soil Pollution, vol. 180, no. 1–4, pp. 199–212, 2007. · ·

41. D. E. Cummings, F. Caccavo Jr., S. Fendorf, and R. F. Rosenzweig, "Arsenic mobilization by the dissimilatory Fe(III)-reducing bacterium Shewanella alga BrY," Environmental Science and Technology, vol. 33, no. 5, pp. 723–729, 1999. · ·

42. C. L. Corkhill and D. J. Vaughan, "Arsenopyrite oxidation—a review," Applied Geochemistry, vol. 24, no. 12, pp. 2342–2361, 2009.

43. K. J. Tufano, C. Reyes, C. W. Saltikov, and S. Fendorf, "Reductive processes controlling arsenic retention: revealing the relative importance of iron and arsenic reduction," Environmental Science and Technology, vol. 42, no. 22, pp. 8283–8289, 2008.

Land Use Suitability Assessment in Low-Slope Hilly Regions under the Impact of Urbanization in Yunnan, China

Gui Jin[1], Zhaohua Li[1], Qiaowen Lin[2], Chenchen Shi[3],
Bing Liu[4], and Lina Yao[5]

[1]Faculty of Resources and Environmental Science, Hubei University, Wuhan, Hubei 430062, China

[2]School of Public Administration, China University of Geosciences, Wuhan 430074, China

[3]State Key Laboratory of Water Environment Simulation, School of Environment, Beijing Normal University, Beijing 100875, China

[4]College of Geomatics, Shandong University of Science and Technology, No. 579 Qianwangang Road, Economic & Technical Development Zone, Qingdao, Shandong 266590, China

[5]Institute of Geographic Sciences and Natural Resources Research, Chinese Academy of Sciences, Chaoyang District, Beijing 100101, China

ABSTRACT

Nowadays, the conflict between land development and land conservation has become increasingly serious in China. The plan called "town of mountain" is carried out in many nonplain areas to alleviate the conflict. To avoid geological disasters and ecological risks in those areas, land use suitability assessment is of great importance. In this paper, the fuzzy weight of evidence model is applied into land use suitability assessment in low-slope hilly regions in Yunnan, China. Fuzzy weight of evidences is calculated to determine 9 map layers. Finally, posterior probabilities are modified after synthesizing each map layer, which are used to generate a land use suitability map. The results show that 9.33%, 26.18%, 45.98%, and 18.51% of low-slope hilly regions are separately highly suitable, moderately suitable, marginally suitable, and unsuitable for development. Besides, highly and moderately suitable areas are mainly located in towns with excellent natural and socioeconomic conditions. The largest areas which are marginally suitable for development are most widely distributed. Unsuitable areas are mainly distributed far away from towns and water sources. The findings of the research will promote the rational use and scientific management of the land.

INTRODUCTION

Due to the rapid urbanization in China [1], the demand for built-up land both in urban and rural areas has been dramatically increased with the shrinking of arable land [2–4]. The conflict between land development and land conservation has become increasingly serious [5–8]. Under the most stringent farmland protection system in China, the strategy of developing low-slope hilly regions is carried out to alleviate the conflict [9]. The concept of low-slope hilly regions is put forward under the demand for land resources in mountain area with the social and economic development and urbanization construction. It refers to contiguously distributed land with slopes less than 25 degrees in the majority of hilly areas as well as unused low mountains and hills which can be used by town construction in the mountain area. It mainly includes a variety of reserved land types such as unused land, abandoned garden, and inefficient forest, which are mostly distributed in the regions with little plain and inadequate protection of cultivated land. Currently, there is a shortage of land available for the development of cities and towns in China [10, 11]. However, low-slope hilly regions account for a relatively large proportion of the available land. Current land use in China, such as a small amount of land per person, intense exploitation of land development, and shortage of reserved land, has limited the development of regional economy and society [12, 13]. Therefore, it is a good way to alleviate the conflict between land and people in the developed regions with high pressure of protecting arable land by optimally using low-slope hilly land [14]. Compared to the plain areas, low-slope hilly land has the lower ecological carrying capacity and higher ecological sensitivity. Lack of good understanding of low-slope hilly land and objective evaluation of development activities will bring enormous environmental disturbance and destruction, resulting in serious or even irreversible consequences on the structure and function of ecosystems, biodiversity, and landscape in this area [15, 16]. Therefore, it is of great significance to do the suitability assessment in low-slope hilly land and classify development levels in order to ensure the orderly development of it [17–19]. Since a framework for land evaluation reported by the Food and Agriculture Organization (FAO), the research of land use suitability assessment receives more and more attention. The theory, method, and specific scheme on land use suitability evaluation are also improved constantly. Currently, scholars usually carry out land suitability assessment in different regions [16, 20–22]. However, the screening of evaluation factor and the determination method of factor weight are both significant on the influence of the cultivated land suitability assessment. On the one hand, the evaluation factor is different in different areas such as the plateau region, hilly area, and flood plain area, and the emphasis of the research is different. Accordingly, on the other hand, the calculation method of factor weight can be divided into two categories [12, 23, 24]: one is the mathematical logical reasoning based on knowledge and rules, containing the comprehensive index method and Fuzzy-AHP methods; this kind of methodology has higher dependency on knowledge and then makes the result more subjective. The other one is the data driven mechanism based on adaptive system such as neural network method. But its reasoning process is cumbersome and its algorithm is complex, which cannot effectively use existing knowledge and often leads to the fact that it is hard to explain the acquired rules. In view of the special natural and social economic conditions in Yunnan region, this paper determines the factor weights by the fuzzy weight of evidence model to avoid subjective influence, multiple collinear interference, and complex model algorithm. The paper provides a scientific basis for urbanization and optimal allocation of land through evaluating the development suitability levels of low-slope hilly resources (not including the ecological preservation areas) such as unused land, woodland, and grassland.

STUDY AREA

Yunnan province (97°31'~106°11'W, 21°8'~29°15'N) is a low-latitude inland, spreading about 394,000 square kilometers which accounts for 4.11 percent of total area in China (Figure 1). It comprises mainly mountains, plateaus, hills, and small basins. Mountains and plateaus occupy around 94% of the province. Much of the province lies within the subtropical highland. There is little variance in annual temperature but a large diurnal temperature range. It has distinct wet and dry seasons. The temperature changes greatly with the terrain. Due to the interaction of climate, biology, geology, topography, and so on, various types of soil are formed. It is characterized by the vertical distribution of soil. The area of red soil accounts for 50% of the province. Rainfall is unevenly distributed in terms of seasons and regions [25]. Yunnan province, with strong intensity in urban land use, is one of the provinces where mountain towns are widespread in China. Therefore, it is an inevitable choice for urbanization development to moderately develop the low-slope hilly regions in Yunnan province.

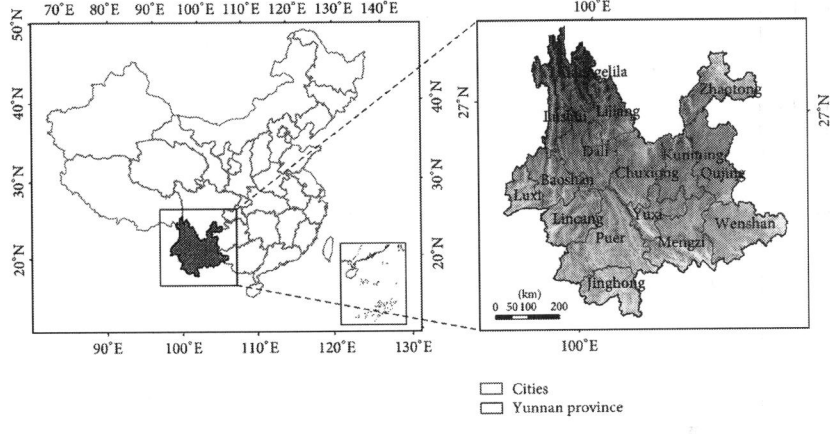

Figure 1: Location and DEM of Yunnan province.

MATERIALS AND METHODS

Data Acquisition

The main data in this study includes meteorological data mainly referring to daily average temperature in 2010 and rainfall in Yunnan province. These data were derived from the daily observation data of meteorological stations maintained by the China Meteorological Administration. Current land use data with the scale 1:100,000 in 2010 of Yunnan province (http://www.resdc.cn) is also used, which contains data regarding all land use types such as built-up land, woodland, grassland, and unused land. Terrain data with the scale 1:25,000 in Yunnan province is from Data Sharing Infrastructure of Earth System Science in China (http://www.geodata.cn/Portal/metadata/viewMetadata.jsp?id=100101-11221). The outcome of overall land use plan (2006–2020) in Yunnan province which derived from the Office of Land and Resources in Yunnan province is used by the research.

ArcGIS10.0 is used to process Kriging interpolation on the rainfall distribution and average temperature distributions with meteorological data, which results in creating an annual rainfall distribution map and an annual temperature distribution map. Topographic data are used to generate digital elevation model (DEM) (whose spatial resolution is 1 km × 1 km) and a slope map. Buffer zones of water sources and state roads are obtained later. Therefore, based on the optimum distance, a map of distance to water and a map of distance to state roads are obtained, respectively. Interpolation analysis is done through socio-economic data, which generates a map of GDP per capita and a population distribution map in Yunnan in the grid with cells of square kilometers [26–28]. The proportion of plain areas is obtained through land use data in Yunnan province [29]. With the help of geoscience data analysis system, the existing built-up land parcels in Yunnan are extracted as training samples suitable for construction and are used to make a distribution map of the sample. It should be noted that GeoDAS4.2 was developed by the Geomatics Research Group at York University, and it is a GIS software that synthesizes the fuzzy weight of evidence model, fractal model, spatial statistics, artificial intelligence, and other modern data processing techniques (http://www.yorku.ca/yul/gazette/past/archive/2002/030602/current.htm). A sampling interval of 1 km × 1 km pixels is determined as the unit of area according to the minimum shape and size of the parcel of evaluation objects. Each map layer is saved in the form of GRID.

Research Methods

Fuzzy weight of evidence is evolved from weight of evidence which is firstly used in medical diagnosis without considering the space. Since 1980s, Agterberg et al. [30, 31] had modified this method and applied it on mineral forecast. Since this method can be applicable to integrate multiple information and spatial decision support systems, it has been applied in the evaluation on various mineral resources, geological disaster risk assessment, and environmental evaluation in recent years.

The Principle of the Method

Weight of evidence method is a log-linear model under a Bayesian probability criterion. A priori probabilities are firstly calculated. Then conditional probabilities are calculated under a certain geological evidence model. Weight of evidence method includes posterior logit model, the general weight of evidence model, fuzzy weight of evidence model, and weighted weight of evidence model. The principle of the method is as follows [32–34].

- *Calculate A Priori Probabilities*: Suppose total area of the study area T is (T) which is divided into cells of a fixed area u. D is the event to be predicted. It follows that there are $N(T) = A(T)/u$ cells totally in the study area where $(\ )$ represents the area and $N(\ )$ is the number of cells. Therefore, the frequency of events D in the study area T is $N(D)$. The probability of the event is

$$P(D) = \frac{N(D)}{T(D)},$$

(1)

Where $P(D)$ is the a priori probability, and the occurrence can be expressed as $O(D) = P(D)/(1 - P(D))$.

- *Calculate Weight of Evidence*: Any weight of evidence (the map layer) x is binary. Its weight is defined as

$$W^{+} = \ln\left[\frac{P(B/D)}{P\left(B/\overline{D}\right)}\right],$$

$$W^{-} = \ln\left[\frac{P\left(\overline{B}/D\right)}{P\left(\overline{B}/\overline{D}\right)}\right],$$

(2)

Where W^{+} and W^{-} represent the weight for presence of B and weight for absence of $\overline{B}$, respectively. $P(B/D) = N(B \cap D)/N(D)$ denotes the probability of selected weight of evidence D in any unit cell of B, where $N(B \cap D)$ stands for the occurrence of weight in B and $N(D)$ is the total number of occurrence in the study area. Other equations can be explained in the same way.

- *Calculate Posterior Probability*: If there is n number of weights which are independent of each other on the occurrence, the probability of the occurrence of any cell in the study area can be expressed by the posterior probability logarithm which is expressed as follows:

$$\ln O\left(D \mid B_1^k \cap B_2^k \cap B_3^k \cdots B_n^k\right) = \sum_{m=1}^{n} W_m^k + \ln O\left(D\right),$$

(3)

Where $m = (1, 2, 3\ldots, n)$, and k indicates the presence and absence of the weight, namely,

$$W_m^k = \begin{cases} W_m^+ \\ W_m^- \\ 0, \end{cases}$$

(4)

Where W_m^+ is the weight for presence of weight m. W_m^- Is the weight for absence of weight m 0 is the weight for missing of weight m accordingly, the posterior probability can be expressed as

$$P = \frac{O}{(1 + O)}.$$

(5)

- *Membership Function*: Weight of evidence approach will result in loss of data while doing binary process for map layers which will affect the accuracy of assessment. Cheng and Agterberg [35] have developed a fuzzy weight of evidence approach which automatically copes with the missing data on discrete layers. Fuzzy weight of evidence uses ambiguity membership of layers which is determined by membership function to deal with multiclassification layers. Multivalued fuzzy membership function $0 \le (x) \le 1$ is used to fit training samples. The weight of evidence is calculated in the end.

Procedures

The procedures of fuzzy weight of evidence (Figure 2) are as follows. (1) Identify research objectives, such as suitability of the development of the low-slope hilly regions; select the samples and calculate the a priori probabilities. (2) Determine the spatial layer related to the target; screen indicators; and establish an index system. (3) Extract map layers related to the target; use fuzzy membership functions to represent the credibility of the map layer. (4) Calculate the weight of fuzzy map layer; determine the value of the map layer; and screen indicators. (5) Integrate multiple fuzzy map layers and calculate a priori probabilities in order to obtain a suitability map in low-slope hilly regions. (6) Modify po

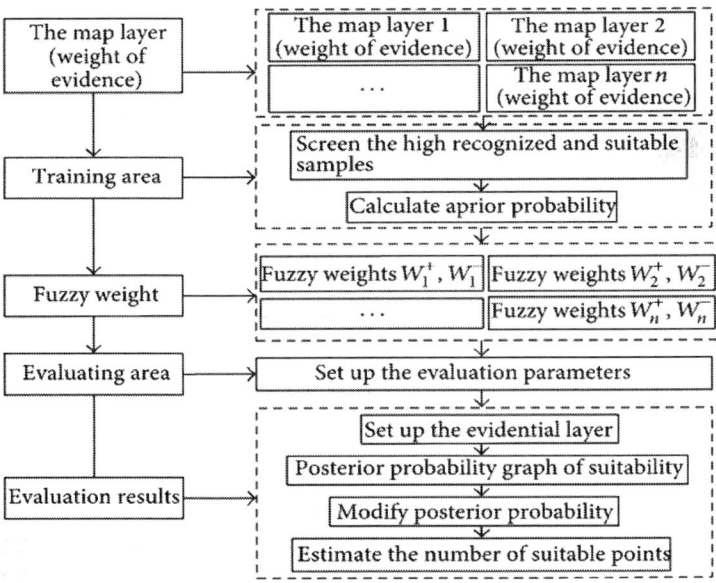

Figure 2: Flowchart of fuzzy weight of evidence.

RESULTS AND ANALYSIS

Training Samples and a Priori Probabilities

That selected training samples should follow two principles. (1) The training samples should be the given information approved by the public and the relevant departments. (2) The training samples should distribute widely in the research area. According to current land use data in Yunnan province, the existing built-up land parcels in Yunnan province are taken as samples suitable for urban development. The priori probability can be known by integrating (1) that the total number of unit cells of training samples is 2332 (Figure 3).There are 220,084 unit cells in the whole areas. The priori probability is 0.010596.

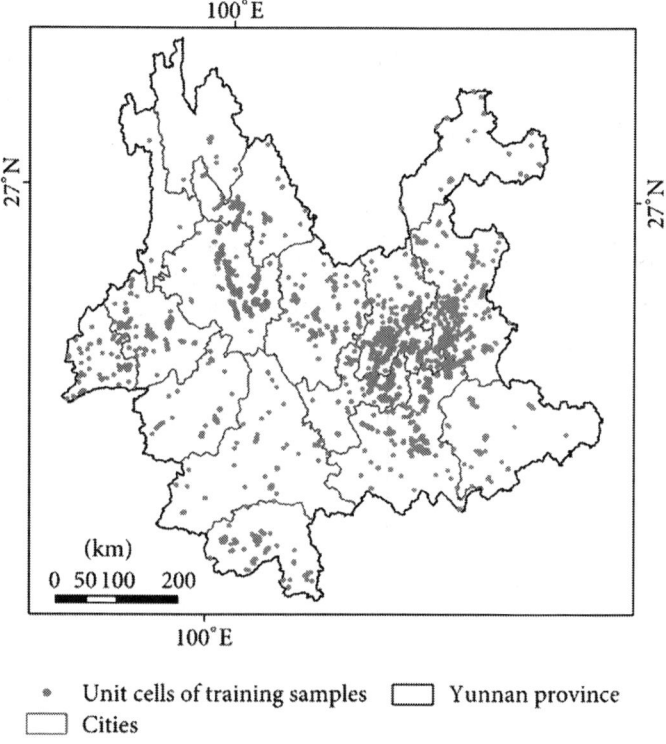

Figure 3: Distribution of training samples.

Fuzzy Map Layers and Weight of Evidence

The selection of evaluation factors is under the principle of dissimilarity, stability, and reality. On the basis of previous research on land use suitability assessment in low-slope hilly regions and built-up land, the evaluation index system of land use suitability assessment in low-slope hilly regions mainly refers to climate, topography, geography, and socioeconomic conditions. Impact factors include temperature, rainfall, elevation, slope, proportion of plain area, distance to water, distance to state roads, population distribution, and GDP per capita (Figure 4).

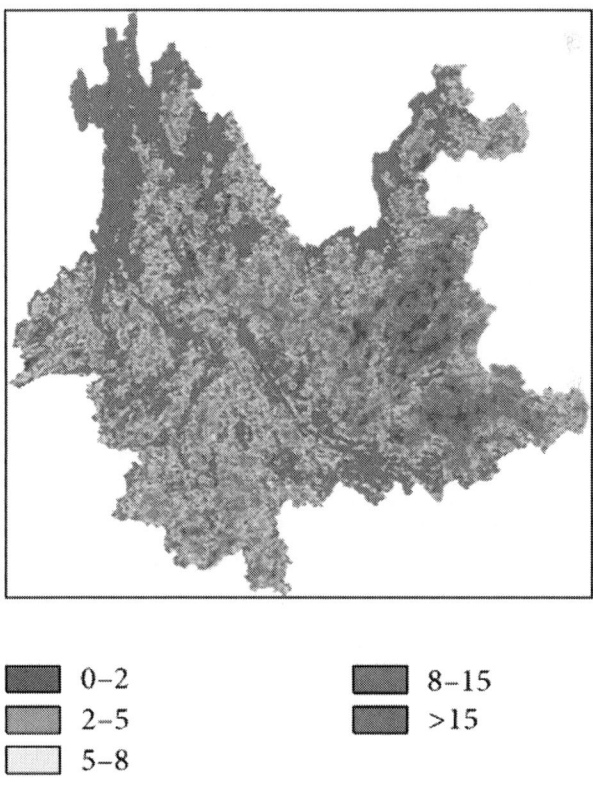

■	0–2	■	8–15
■	2–5	■	>15
□	5–8		

(a)

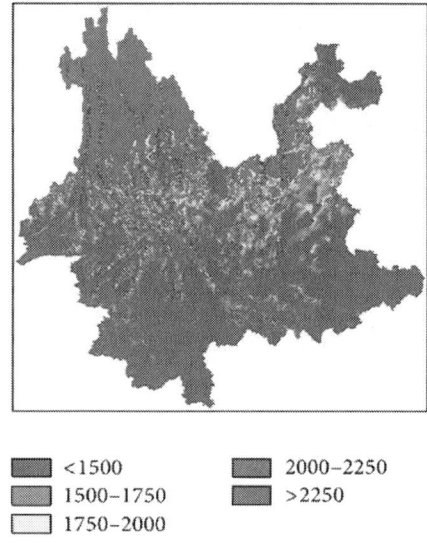

▨ <1500	▨ 2000–2250
▨ 1500–1750	▨ >2250
☐ 1750–2000	

(b)

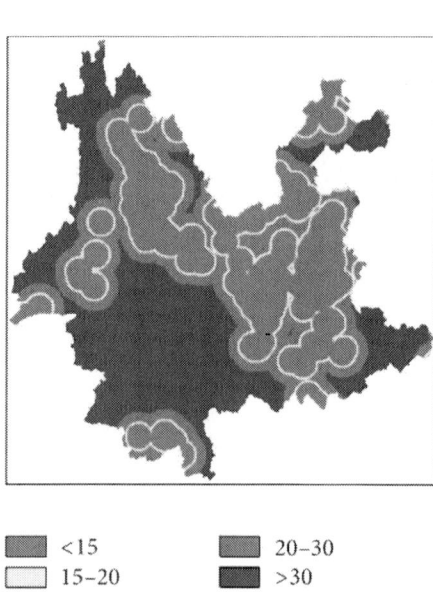

▨ <15	▨ 20–30
☐ 15–20	▨ >30

(c)

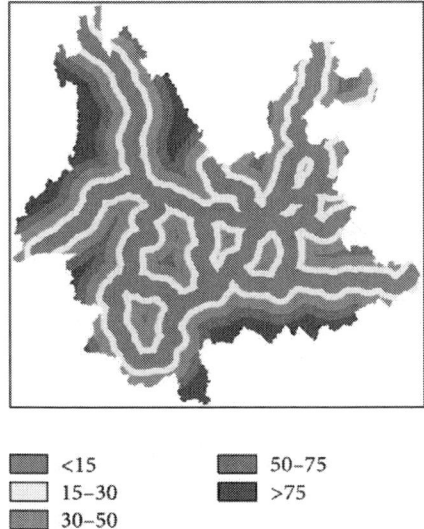

■	<15	■	50–75
□	15–30	■	>75
▨	30–50		

(d)

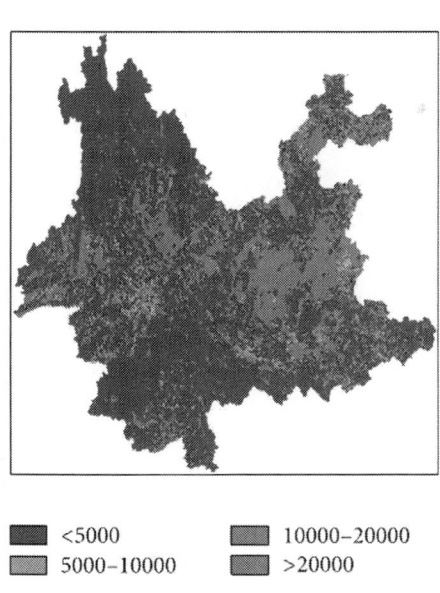

■	<5000	▨	10000–20000
▨	5000–10000	▨	>20000

(e)

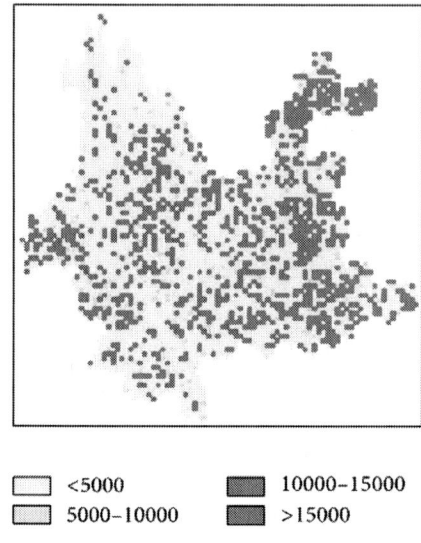

	<5000		10000–15000
	5000–10000		>15000

(f)

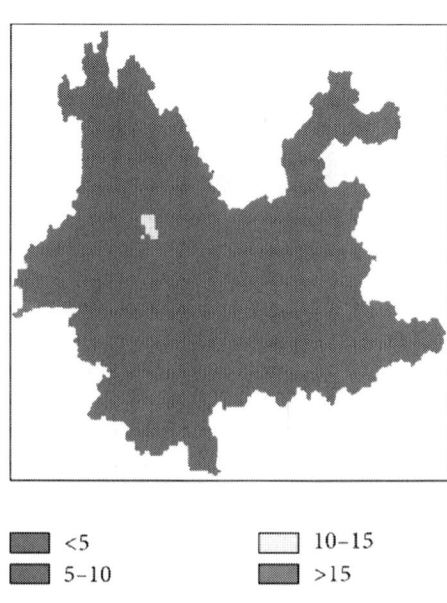

	<5		10–15
	5–10		>15

(g)

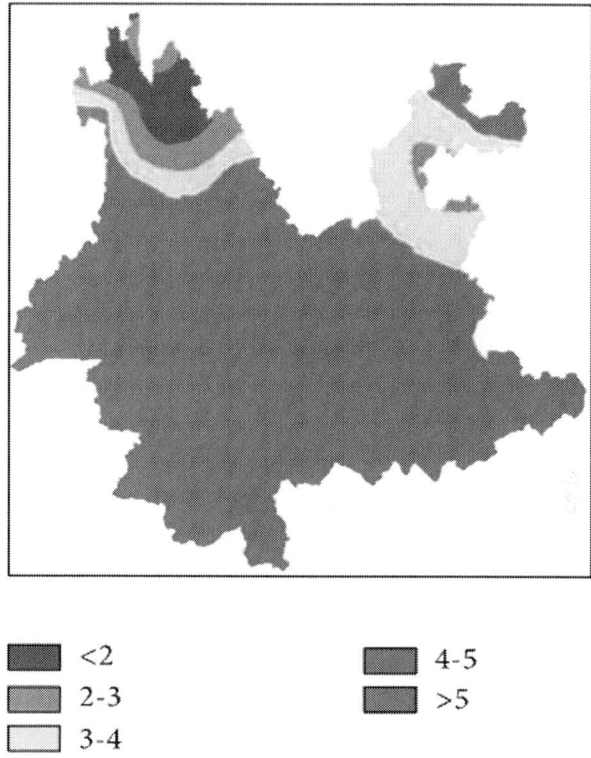

▆ <2		▆ 4-5
▆ 2-3		▆ >5
☐ 3-4		

(h)

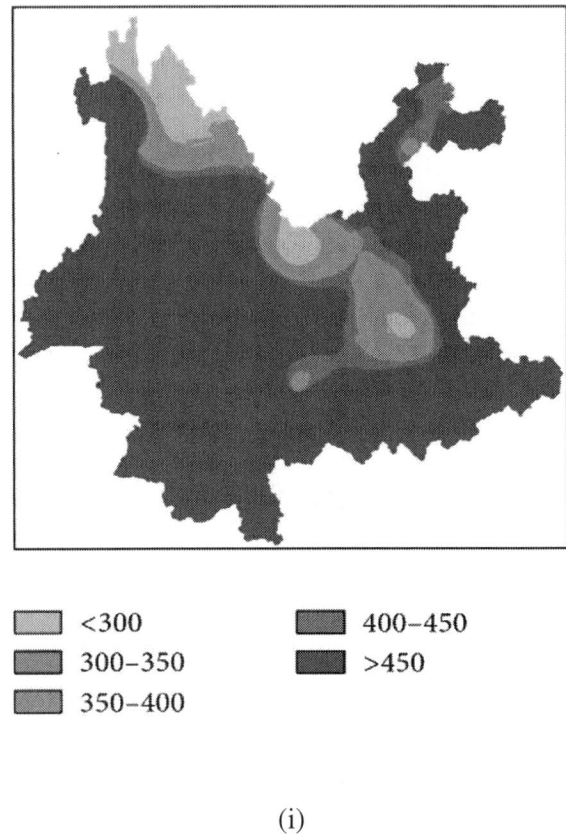

□ <300	■ 400–450
■ 300–350	■ >450
■ 350–400	

(i)

Figure 4: Nine-factor maps. (a) Slope (°); (b) elevation (m); (c) distance to water (Km); (d) distance to state roads (Km); (e) GDP per capita (Yuan); (f) population (person); (g) proportion of plain areas (%); (h) annual average temperature (°C); (i) annual rainfall (mm).

Each fuzzy weight of evidence is calculated by GeoDAS4.2. The influence of individual factors on the incident can be known which leads to the objective decision on choosing evaluation factors. Final map layers and parameters of fuzzy weights of evidence can be seen in Table 1 through (2) and multivalued fuzzy membership function.

Table 1: Parameters of fuzzy weights of evidence for each map layer

	Classification value	+	−		STDEV		MSF	FW	(FW)	(FW)
Slope/(°)	0~2	1.38	−0.69	2.07	0.04	49.10	1.00	1.38	0.03	49.62
	2~5	0.68	−1.21	1.89	0.05	34.35	0.60	1.03	0.08	13.50
	5~8	0.36	−1.44	1.79	0.07	24.30	0.38	0.68	0.12	5.91
	8~15	0.09	−2.03	2.12	0.19	11.32	0.15	−0.06	0.16	−0.37
	>15						0.00	−2.03	0.19	−10.92
Elevation (m)	<1500	0.28	−0.82	1.10	0.06	19.37	1.00	0.28	0.02	12.18
	1500~1750	0.24	−1.15	1.38	0.07	18.99	0.80	0.22	0.03	6.28
	1750~2000	0.20	−1.69	1.89	0.11	17.52	0.68	0.17	0.05	3.40
	2000~2250	0.17	−2.16	2.32	0.15	15.59	0.40	−0.03	0.09	−0.29
	>2250						0.00	−2.16	0.15	−14.62
Distance to water/km	<15	0.83	−1.17	2.00	0.05	38.85	1.0	0.83	0.02	35.72
	15~20	0.71	−1.26	1.97	0.06	35.39	0.8	0.61	0.02	26.25
	20~30	0.41	−1.49	1.90	0.07	26.06	0.4	−0.01	0.04	−0.19
	>30						0.0	−1.49	0.07	−21.43
Distance to state roads/km	<15	0.72	−0.79	1.51	0.05	33.35	1.00	0.72	0.02	29.02
	15~30	0.42	−1.20	1.62	0.06	26.05	0.68	0.63	0.06	11.12
	30~50	0.18	−1.31	1.49	0.09	16.66	0.40	0.46	0.10	4.51
	50~75	0.07	−1.64	1.71	0.17	10.02	0.20	0.16	0.14	1.15
	>75						0.00	−1.64	0.17	−9.71

GDP per capita/Yuan	<5000						0.00	-2.24	0.08	-28.60
	5000~10000	1.00	-2.24	3.24	0.08	39.85	0.40	-0.19	0.05	-3.92
	10000~20000	1.31	-1.97	3.28	0.07	48.71	0.85	1.10	0.02	48.67
	>20000	1.60	-1.74	3.34	0.06	56.64	1.00	1.60	0.02	70.40
Population/person	<5000	0.01	-0.77	0.78	0.22	3.55	0.0	-0.38	0.03	-12.26
	5000~10000	0.52	-0.39	0.91	0.04	21.71	0.6	0.11	0.02	5.18
	10000~15000	0.58	-0.36	0.94	0.04	22.62	0.8	0.35	0.03	13.78
	>15000	0.65	-0.34	0.99	0.04	23.84	1.0	0.65	0.03	21.21
Proportion of plain areas/%	<5	0.00	-1.38	1.38	0.58	2.38	0.0	-0.04	0.02	-1.74
	5~10	0.59	-0.04	0.64	0.07	8.60	0.4	-0.02	0.05	-0.39
	10~15	1.77	-0.03	1.80	0.12	14.74	1.0	1.77	0.12	14.75
	>15	0.80	0.00	0.81	0.34	2.40	1.0	1.77	0.12	14.75
Annual average temperature/°C	<2						0.00	-2.16	0.30	-7.18
	2~3	0.04	-2.17	2.20	0.30	7.28	0.65	0.05	0.11	0.46
	3~4	0.07	-1.85	1.92	0.19	10.29	1.00	0.07	0.02	3.36
	4~5	0.12	-0.86	0.98	0.08	12.55	1.00	0.07	0.02	3.36
	>5	-0.33	0.44	-0.77	0.04	-18.62	1.00	0.07	0.02	3.36
Annual rainfall/mm	<300						0.0	-1.40	0.27	-5.25
	300~350						0.0	-1.40	0.27	-5.25
	350~400	0.02	-1.40	1.42	0.27	5.31	0.3	-0.09	0.19	-0.50
	400~450	0.02	-0.47	0.49	0.11	4.39	0.8	-0.05	0.06	-0.91
	>500	-0.05	0.23	-0.27	0.05	-5.16	1.0	-0.05	0.02	-2.00

W^+ : positive weight; W^- : negative weight; C: contrast of W^+ - W^- ; STDEV: standard deviation of C; T: significance level; MSF: membership; FW: fuzzy weight; S(FW): standard deviation of fuzzy weight; T(FW): significance of fuzzy weight. Among them, GDP is the present value of 2010.

The Posterior Probability and Its Modification

These map layers are synthesized and the posterior probability map in Yunnan is calculated by (5). Then the posterior probability map of land use suitability in low-slope hilly regions in Yunnan is made by taking reserved land parcels in low-slope hilly regions as the crop box (Figure 5). Posterior probabilities are affected by the setting of the unit cell. However, they do not affect the distribution of posterior probabilities. Therefore, posterior probabilities do not represent the probability of occurrences but the distribution after occurrences.

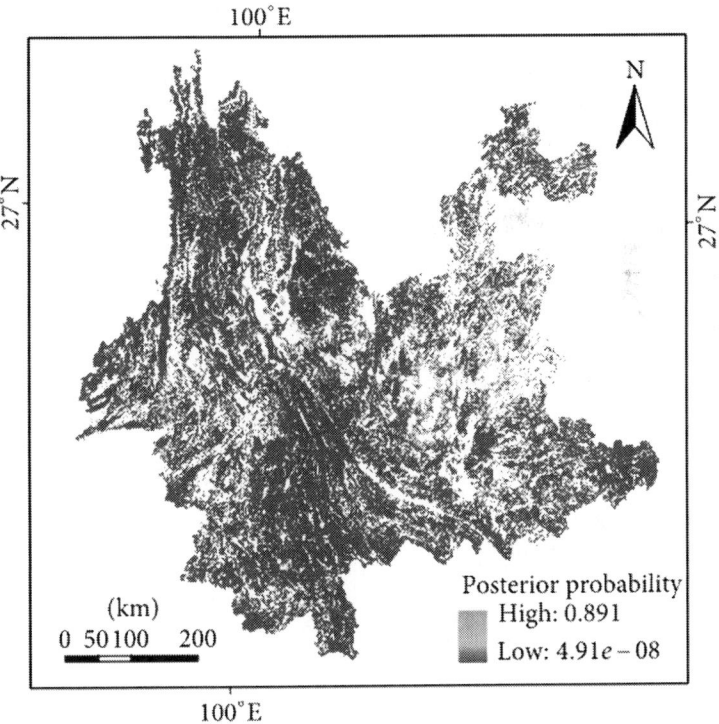

Figure 5: Posterior probability map of land use suitability.

The map layer of fuzzy weight of evidence model is required to meet the needs of conditional independence. However, in reality, it is very difficult to achieve this which inevitably results in the deviation of posterior probabilities. The modified posterior probability can overcome the deficit of less accurate estimation on appropriate points.

Data tests of the linear function correction, logarithmic function correction, and exponential function correction are carried out with the help of GeoDAS4.2's posterior probability correction module. Ultimately, exponential function model $Y = 0.12x^{0.96}$ with with optimal fitting degree is determined. After the modification of the posterior probability, the coefficient of determination R is equal to 0.98. Procedures and results of the modification are shown in Figures 6 and 7.

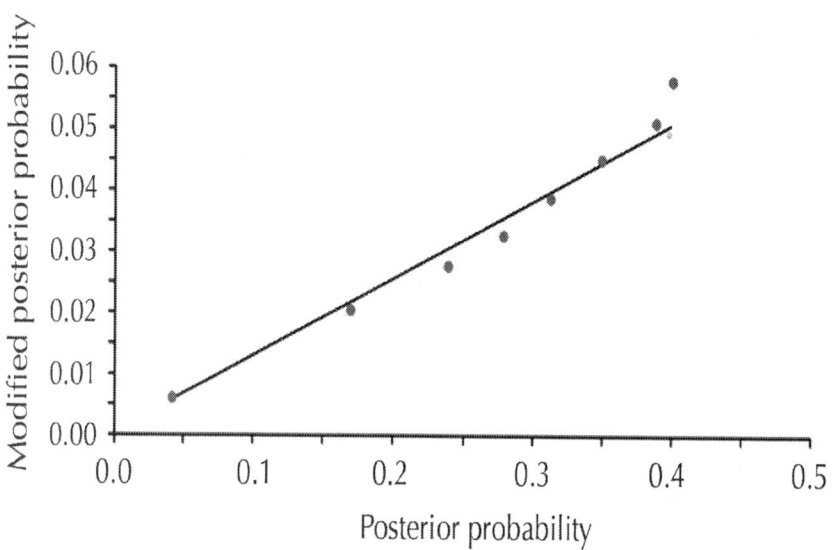

Figure 6: Modification on posterior probabilities.

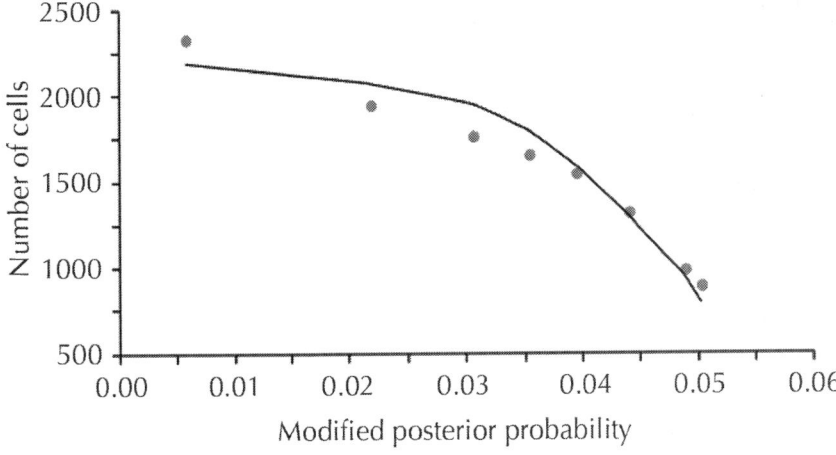

Figure 7: Modified posterior probabilities of land use suitability and numbers of corresponding points.

Classification of Land Use Suitability in Low-Slope Hilly Regions

The posterior probability map is modified according to an exponential function model, which is used to create the modified posterior probability map. On the basis of this, the distribution frequency of modified posterior probabilities is analyzed. Distribution frequency curve is used. Finally, land use suitability in low-slope hilly regions in Yunnan province is divided into four levels (Table 2, Figure 8) according to obvious inflection point of the frequency curve and parameters of the weight of evidence of those evaluation factors.

Table 2: Classification of land use suitability in low-slope hilly regions in Yunnan

Modified posterior probability	Suitability level	Numbers of cells	Description	Percentage of area/%
>0.01047~0.10860	Level 1	20315	Highly suitable	9.33

>0.00050~0.10467	Level 2	56997	Moderately suitable	26.18
>0.00002~0.00050	Level 3	100131	Marginally suitable	45.98
<0.00002	Level 4	40309	Unsuitable	18.51

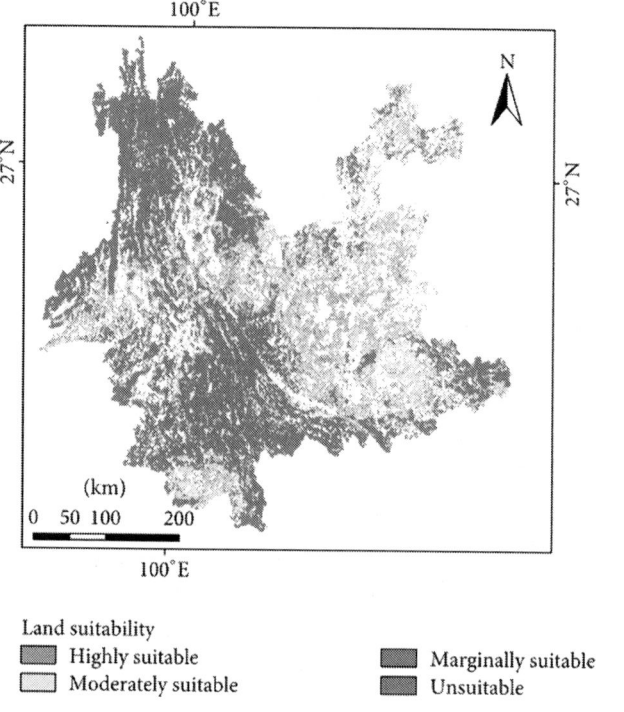

Figure 8: Classification map of land use suitability in low-slope hilly regions in Yunnan.

There are four levels in terms of land use suitability in low-slope hilly regions in the study area: highly suitable, moderately suitable, marginally suitable, and unsuitable. (1) 9.33% of the land is highly suitable for development, mostly distributed in Kunming, Lijiang, Dali Bai autonomous prefecture, Qujing, and so on. Spatially, highly suitable areas are mainly located around cities where its terrain is relatively flat with slopes below 5 degrees and altitude mostly below 1500 m. The plain area accounts for more than 15% of the total area.

Usually, the distance to water and state roads is no more than 15 km. GDP per capita is higher than 20,000 yuan. It is densely populated with 1,500 people/km². Its average annual temperatures are relatively high with abundant rainfall. (2) 26.18% of land is moderately suitable where it still has good natural resource and socioeconomic conditions with slopes ranging from 5 degrees to 8 degrees. It is close to towns. GDP per capita ranging from about 10,000 to 20,000 yuan is high. The population is relatively concentrated in these areas with a population density of 10,000 to 15,000 people/km². It has adequate rainfall and pleasant weather suitable for living. (3) 45.98% of land is marginally suitable for development. It is widely spread. Compared with highly suitable and moderately suitable areas, basic conditions in this area are relatively poor. However, it has strong potential for transformation, which can be the main source of future development for some time. Therefore, its development potential, the overall size, and spatial layout should be the focus of policy makers. (4) 18.51% of land is unsuitable for development. These areas are more dispersed with slopes above 15 degrees in general and altitude mostly above 2200 m. The proportion of plain area is less than 5% of the total area. It is far away from water resources and state roads. GDP per capita is low and population density is relatively small. It is far away from cities and towns. It is not suitable for living due to the relatively unpleasant weather. At the same time, it is prone to geological disasters and ecological risks, inappropriate for development and construction. In summary, the four levels of the suitability assessment in low-slope hilly regions in Yunnan province can provide a basis for the priority of development in these regions. Under the consideration of social-economic development, urbanization, and ecological safety, the paper offers strategies to develop low-slope hilly regions, which focus on the exploration of unused land and give preferential protection to woodland and grassland. In addition, comparing the results of this paper with the achievement of the outcome of overall land use plan, it can be known that, as far as quantity structure is concerned, it is quite scientific and reasonable that the area of highly suitable resources of low-slope hilly regions is much larger than the built-up land index of the planning phase (2006–2020); as for the spatial distribution, the layout of the built-up land in the planning phase is basically consistent with the highly suitable low-slope hilly regions or is distributed in its perimeter zone.

CONCLUSIONS AND DISCUSSION

Due to global climate change and urbanization in China, the conflict between land development and land conservation has become increasingly serious [36]. The development of low-slope hilly regions is an important measure to alleviate the conflict. This paper takes Yunnan province as a typical case study, integrating natural geographical features and elements of human geography in Yunnan province to analyze land use suitability of low-slope hilly regions in Yunnan province and avoid potential geological disasters and ecological risks in those regions. Land use suitability of low-slope hilly regions in Yunnan province is evaluated by analyzing some factors referring to climate, topography, geography, society, and economy. Nine map layers regarding temperature, rainfall, elevation, slope, proportion of plain areas, distance to water, distance to state roads, population distribution, and GDP per capita are selected. The results show that (1) 9.33% of low-slope hilly regions in Yunnan are highly suitable for development. 26.18% of land is moderately suitable. 45.98% of land is generally suitable. The remaining 18.51% of land is unsuitable for development. This outcome of spatial distribution of low-slope hilly land in Yunnan province can finally optimize the size and layout of low-slope hilly land and promote the rational use of it. However, it is worth mentioning that not only the suitability of natural quality should be considered, but also some factors such as social and economic suitability, ecological suitability, and land policy should be taken into consideration. The related research will provide references for the plan of "town of mountain" in Yunnan and for the development of low-slope hilly land in Yunnan province. As for the model used in this study, it should be pointed out that training samples chosen in the study must be highly recognized appropriate points to ensure scientific results. Therefore, future research should pay more attention to how to ensure diversification of the given sources and improve the accuracy of given information.

ACKNOWLEDGMENTS

This research was financially supported by the major research plan of the National Natural Science Foundation of China (Grant no. 91325302), the National Natural Science Funds of China for Distinguished Young

Scholar (Grant no. 71225005), and National Key Programme for Developing Basic Science in China (Grant no. 2010CB950900).

REFERENCES

1. R. Lal, "Soil carbon sequestration impacts on global climate change and food security," Science, vol. 304, no. 5677, pp. 1623–1627, 2004. · ·

2. G. C. S. Lin and S. P. S. Ho, "China›s land resources and land-use change: insights from the 1996 land survey," Land Use Policy, vol. 20, no. 2, pp. 87–107, 2003. · ·

3. Y. Xie, Y. Mei, T. Guangjin, and X. Xuerong, "Socio-economic driving forces of arable land conversion: a case study of Wuxian City, China," Global Environmental Change, vol. 15, no. 3, pp. 238–252, 2005. · ·

4. P. Hao, S. Geertman, P. Hooimeijer, and R. Sliuzas, "The land-use diversity in urban villages in Shenzhen," Environment and Planning A, vol. 44, no. 11, pp. 2742–2764, 2012. · ·

5. L. Jiang, X. Deng, and K. C. Seto, "Multi-level modeling of urban expansion and cultivated land conversion for urban hotspot counties in China," Landscape and Urban Planning, vol. 108, no. 2–4, pp. 131–139, 2012. · ·

6. L. Jiang, X. Deng, and K. C. Seto, "The impact of urban expansion on agricultural land use intensity in China," Land Use Policy, vol. 35, pp. 33–39, 2013. · ·

7. E. Keys and W. J. McConnell, "Global change and the intensification of agriculture in the tropics,"Global Environmental Change, vol. 15, no. 4, pp. 320–337, 2005. ·

8. I. Eng, "The rise of manufacturing towns: externally driven industrialization and urban development in the Pearl River Delta of China," International Journal of Urban and Regional Research, vol. 21, no. 4, pp. 554–568, 1997. · ·

9. Y. Liu and Y. Li, "Environment: China›s land creation project stands firm," Nature, vol. 511, no. 7510, article 410, 2014. ·

10. J. Liu, J. Zhan, and X. Deng, "Spatio-temporal patterns and driving forces of urban land expansion in China during the economic reform era," Ambio, vol. 34, no. 6, pp. 450–455, 2005.

11. X. Deng, J. Huang, S. Rozelle, and E. Uchida, "Economic growth and the expansion of urban land in China," Urban Studies, vol. 47, no. 4, pp. 813–843, 2010. ··

12. J. Gong, Y. Liu, and W. Chen, "Land suitability evaluation for development using a matter-element model: a case study in Zengcheng, Guangzhou, China," Land Use Policy, vol. 29, no. 2, pp. 464–472, 2012. ··

13. J. S. Jeong, L. García-Moruno, and J. Hernández-Blanco, "A site planning approach for rural buildings into a landscape using a spatial multi-criteria decision analysis methodology," Land Use Policy, vol. 32, pp. 108–118, 2013. ··

14. P. La Greca, D. La Rosa, F. Martinico, and R. Privitera, "Agricultural and green infrastructures: the role of non-urbanised areas for eco-sustainable planning in a metropolitan region," Environmental Pollution, vol. 159, no. 8-9, pp. 2193–2202, 2011. ··

15. R. S. Sicat, E. J. M. Carranza, and U. B. Nidumolu, "Fuzzy modeling of farmers› knowledge for land suitability classification," Agricultural Systems, vol. 83, no. 1, pp. 49–75, 2005. ··

16. T. V. Reshmidevi, T. I. Eldho, and R. Jana, "A GIS-integrated fuzzy rule-based inference system for land suitability evaluation in agricultural watersheds," Agricultural Systems, vol. 101, no. 1-2, pp. 101–109, 2009. ··

17. J. Malczewski, "Ordered weighted averaging with fuzzy quantifiers: GIS-based multicriteria evaluation for land-use suitability analysis," International Journal of Applied Earth Observation and Geoinformation, vol. 8, no. 4, pp. 270–277, 2006. ·

18. J. Marull, J. Pino, J. M. Mallarach, and M. J. Cordobilla, "A land suitability index for strategic environmental assessment in metropolitan areas," Landscape and Urban Planning, vol. 81, no. 3, pp. 200–212, 2007. ··

19. R. J. Zomer, A. Trabucco, D. A. Bossio, and L. V. Verchot, "Climate change mitigation: a spatial analysis of global land suitability for clean development mechanism afforestation and reforestation," Agriculture, Ecosystems and Environment, vol. 126, no. 1-2, pp. 67–80, 2008. ··

20. S. V. Bobade, B. P. Bhaskar, M. S. Gaikwad et al., "A GIS-based land use suitability assessment in Seoni district, Madhya Pradesh, India," Tropical Ecology, vol. 51, no. 1, pp. 41–54, 2010.

21. D. D:haeze, J. Deckers, D. Raes, T. A. Phong, and H. V. Loi, "Environmental and socio-economic impacts of institutional reforms on the agricultural sector of Vietnam: land suitability assessment for Robusta coffee in the Dak Gan region," Agriculture, Ecosystems & Environment, vol. 105, no. 1-2, pp. 59–76, 2005. · ·

22. Y.-S. Liu, J.-Y. Wang, and L.-Y. Guo, "GIS-based assessment of land suitability for optimal allocation in the Qinling Mountains, China," Pedosphere, vol. 16, no. 5, pp. 579–586, 2006. · ·

23. Y. Chen, J. Yu, and S. Khan, "Spatial sensitivity analysis of multi-criteria weights in GIS-based land suitability evaluation," Environmental Modelling and Software, vol. 25, no. 12, pp. 1582–1591, 2010. · ·

24. O. Marinoni, "A discussion on the computational limitations of outranking methods for land-use suitability assessment," International Journal of Geographical Information Science, vol. 20, no. 1, pp. 69–87, 2006. · ·

25. C. Zhao, X. Deng, Y. Yuan, H. Yan, and H. Liang, "Prediction of drought risk based on the WRF model in Yunnan province of China," Advances in Meteorology, vol. 2013, Article ID 295856, 9 pages, 2013. · ·

26. X. Deng, H. Su, and J. Zhan, "Integration of multiple data sources to simulate the dynamics of land systems," Sensors, vol. 8, no. 2, pp. 620–634, 2008. · ·

27. X. Deng, J. Huang, Q. Huang, S. Rozelle, and J. Gibson, "Do roads lead to grassland degradation or restoration? A case study in Inner Mongolia, China," Environment and Development Economics, vol. 16, no. 6, pp. 751–773, 2011. · ·

28. X. Deng, J. Huang, E. Uchida, S. Rozelle, and J. Gibson, "Pressure cookers or pressure valves: do roads lead to deforestation in China?" Journal of Environmental Economics and Management, vol. 61, no. 1, pp. 79–94, 2011. · ·

29. X. Deng, J. Huang, S. Rozelle, and E. Uchida, "Growth, population and industrialization, and urban land expansion of

China," Journal of Urban Economics, vol. 63, no. 1, pp. 96–115, 2008. · ·

30. F. P. Agterberg, "Computer programs for mineral exploration," Science, vol. 245, no. 4913, pp. 76–81, 1989. · ·

31. F. Agterberg, G. Bonham-Carter, Q. Cheng, and D. Wright, "Weights of evidence modeling and weighted logistic regression for mineral potential mapping," Computers in Geology, vol. 25, pp. 13–32, 1993.

32. F. P. Agterberg and Q. Cheng, "Conditional independence test for weights-of-evidence modeling," Natural Resources Research, vol. 11, no. 4, pp. 249–255, 2002.

33. F. Agterberg, "A modified weights-of-evidence method for regional mineral resource estimation," Natural Resources Research, vol. 20, no. 2, pp. 95–101, 2011. · ·

34. G. Jin, Z. Wang, X. Hu, S. Hu, and D. Zhang, "Land suitability evaluation in Qinghai-Tibet Plateau based on fuzzy weight of evidence model," Transactions of the Chinese Society of Agricultural Engineering, vol. 29, no. 18, pp. 241–250, 2013 (Chinese). ·

35. Q. Cheng and F. P. Agterberg, "Fuzzy weights of evidence method and its application in mineral potential mapping," Natural Resources Research, vol. 8, no. 1, pp. 27–35, 1999. · ·

36. J. Luo, J. Zhan, Y. Lin, and C. Zhao, "An equilibrium analysis of the land use structure in the Yunnan Province, China," Frontiers of Earth Science, 2014. · ·

The Analysis of Pricing Power of Preponderant Metal Mineral Resources under the Perspective of Intergenerational Equity and Social Preferences: An Analytical Framework Based on Cournot Equilibrium Model

Meirui Zhong[1, 2], Anqi Zeng[1], Jianbai Huang[1, 2], and Jinyu Chen[1, 2]

[1]School of Business, Central South University, Changsha 410083, China

²Institute of Metal Resources Strategy, Changsha 410083, China

ABSTRACT

This paper combines intergenerational equity equilibrium and social preferences equilibrium with Cournot equilibrium solving the technological problem of intergenerational equity and strategic value compensation confirmation, achieving the effective combination between sustainable development concept and value evaluation, thinking and expanding the theoretical framework for the lack of pricing power of mineral resources. The conclusion of the theoretical model and the numerical simulation shows that intergenerational equity equilibrium and social preferences equilibrium enhance international trade market power of preponderant metal mineral resources owing to the production of intergenerational equity compensation value and strategic value. However, the impact exerted on Cournot market power by social preferences is inconsistent: that is, changes of altruistic Cournot equilibrium and reciprocal inequity Cournot equilibrium are consistent, while inequity aversion Cournot equilibrium has the characteristic of loss aversion, namely, under the consideration of inequity aversion Cournot competition, Counot-Nash equilibrium transforms monotonically with sympathy and jealousy of inequity aversion.

INTRODUCTION

Pricing power is the ability where related market participants manipulate market equilibrium price away from international trade fair price in its favor by market forces. In recent years, the sustainable growth of China's economy is the important engine driving the growth of the world economy and the increasing demand of staple commodities, such as metal mineral products. The influence exerted on global economy by China is called "China Factor" internationally. However the so-called "China Factor" does not bring corresponding pricing power to China; instead, metal mineral resources international trade price of our country is stuck in the dilemma. The export price of preponderant metal mineral resources, such as rare earth, lithium, and indium, experienced a long-term slump, which not only cause economic loss but also leave the burden of energy consumption and environmental protection to China, so that it is equivalent to providing hidden subsidies at the cost of ecological environment destruction and mineral resources rapid depletion. Therefore, the reports of the seventeenth and eighteenth congress of the CPC put forward continuously [1, 2], "deepen the resource products price and tax reform, establish compensated use system and eco-compensation system reflected the market supply and demand, resource scarcity and intergenerational compensation." Although intergenerational compensation is stressed in reports, externalities resulting from productive process of metal mineral resources development and utilization are not included in metal mineral resources value system as the form of cost. The proportion of calculated mineral resources compensation fees to sales revenue is approximately 1.18%, far lower than the level of 2%–8% of foreign premium. The premium rate and compensation fees of rare earth in China are far lower than Australia, the USA, South Africa, and Vietnam. The technological problem of cost confirmation and measurement and serious distortion of tax policy prevent mineral resource development and utilization from reasonable value compensations, causing unfairness in international trade fair price, and above all that is the important reason why China loses pricing power.

Considering that the international trade price of metal mineral resources is also affected by factors such as supply and demand and speculation (expectation), the complete value compensation system including marginal cost of production, marginal user cost, and external cost (ecological value and intergenerational compensation value) is only the static reason to explain pricing power deficiency. Especially to preponderant metal mineral resources, such as rare earth, lithium, and indium, their international trade price is mainly dependent on mutual bargaining, which is affected by psychological preferences of players and thus produce strategic value. While ignoring objectivity of strategic value in policy suggestion making is another reason accounted for pricing power deficiency. Because according to social preference equilibrium analysis, unless the international trade price is fair and it achieves equilibrium among players, it is difficult for metal mineral resources development and utilization to achieve success. Using traditional game equilibrium evaluation method can reflect the economic value connotation accurately; however, a certain mineral resource development and utilization are accepted only when players approve of economic value and ecological value and the fairness of metal mineral resources development compensation price psychologically. So fairness correction on metal mineral resources development compensation basic value is necessary in reality.

Classical literatures discussing pricing fairness from a perspective of mineral resources development value compensation include "Hotelling Rule" [3], raised by Hotelling, namely, mineral resources, as a kind of asset, need depreciation, so the depletion of mineral resources could be compensated by taxation; "Hartwick Rule" [4] raised by Hartwick, pointed out that if mining rent of nonrenewable resources saved as productive investment, the investment across generations is equal to achieve sustainability of economic growth, when the investment is greater than resources value extracted by resources owners. Serafy [5] adds environmental losses to the research work of national income accounting system and raises user cost approach to calculate real income and the value depletion of nonrenewable resources, which is a new national income accounting method in nonrenewable resources field; later, Serafy [6] makes improvement in this method. This approach lays the foundation of depletion cost pricing of nonrenewable mineral resource; therefore, it is used by many scholars to measure user cost of various mineral resources and analyze the reasonability of premium system and resources tax and fee policy, such as Adelman [7] calculates user cost of some large oil and gas companies by user cost approach and compares with premium; Young and Seroa Da Motta [8] count user cost of major minerals in Brazil by this method; Blignaut and Hassan [9] estimate user costs of underground mineral resources in South Africa; Lin et al. [10] discover the inadaptability of user cost approach in coal resources of China; thus, he uses the modified approach to estimate real cost of coal resources and builds CGE model to determine detailed tax rates; G. P. Li and H. W. Li [11] correct the defects of user cost method and use it to calculate user cost of oil and gas in the United States; Zeng and Li [12] use fixed user cost approach to count user cost of coal, oil, and natural gas in China during 1985–2010, after taking depletion in resource development and the effects of inflation into account. These researches above solve problems of metal mineral resources compensation scarcity value but ignore environmental costs and intergenerational equity value and lack explanation for influences exerted on market and international trade price by intergenerational equity compensation value. On the other hand, many scholars do researches on pricing power, such as Fattouh [13] who suggests that pricing power is the ability for manufacturers directly affecting the other market participants and market variables, such as price and sales, so market pricing power is a kind of price bonus ability; Kaufmann [14] argues that pricing power is the technical strength which is associated with market power to some extent, that is, enterprises could obtain monopoly pricing power in the market by its unique technology or patents, thus gaining excess profits. Rubinstein [15] explains staple commodity pricing mechanism by using the bargaining model of complete information dynamic game and deems that pricing power advantage between buyers and sellers mainly depends on bargaining patience of two sides when information is complete. Wen et al. [16–19] hold that influences on market structure carried by risk preference and risk premium should be taken into account in bargaining model, for risk preference characteristics will affect pricing power by affecting market power. Some researches specific to rare earth pricing power following the above trend are carried out, such as Zhang [20] who thinks that the rare earth market belongs to a typical oligopoly market, so oligopolistic enterprise must fully consider impacts from competitors before taking any action, which proves game behavior of oligopolists on both sides existed in the pricing process of rare earth; Wang and Zhang [21] analyze the potential impact on China's rare earth export pricing power by the increase of resource tax; Wu and Jiang [22] hold that the formation of pricing power is a result of comprehensive shaping process involving many factors, such as, industry, enterprise, government, and foreign aspects, which all are passed on to the market power.

The analysis above is static interpretation of pricing power, without considering the influence on market power by psychological preferences. Based on remarkable discovery of game experiment, behavioral economics expand and correct the traditional economic theory through integrating behavioral and psychological preferences into it, especially blending social preference in game and decision-making theories. As an effective analytical tool for economic subject of cooperative game, it brings profound impact on the raise of fairness preference and application in motivation theory and industrial organizational theory, such as Rabin [23] who starts original research toward fair game equilibrium and builds a reciprocal fairness equilibrium game model based on the framework of psychological game raised by Geanakoplos, Pearce, and Stacchetti. This model depicts reciprocal fairness motivation of players as motivation fairness utility function and then discovers a new equilibrium, that is, fairness equilibrium, which meets the Pareto optimality with cooperative equilibrium and provides a reasonable explanation for cooperative results. However, Rabin's model is difficult to predict accurately because it only aimed at games with standard form, not for dynamic game with continuous strategy structure. Dufwenberg and Kirchsteiger [24] improve Rabin's model through expanding it to a dynamic environment with continuous strategy structure, thus obtaining a more extensive application. According to fairness preference based on distribution results revealed by game experiments, Fehr and Schmidt [25] and Bolton and Ockenfels [26] develop inequity aversion model based on distribution results. It could be deduced from research achievements of reciprocal equity equilibrium theory that metal mineral resources development and compensation value are dependent on not only material benefits brought by resource development, but also psychological effect contained by reciprocal fairness belief, if only a reciprocal fairness belief of related subjects is given. Therefore, a correlation consideration is established between reciprocal equity equilibrium analysis and metal mineral resources development and compensation evaluation, consequently revealing the mechanism of pricing power affected by psychological preferences. The breakthrough of theory model in psychological preferences utility assumes the analysis of pricing power affected by market power ignoring psychological preferences, which will affect the bargaining strategy of players, change the market power and influence supply and demand prices of mineral products. For instance, Zhong et al. [27] estimate intergenerational compensation of preponderant high-tech mental mineral resources affected by altruism preference and reciprocal fairness equilibrium with Stackelberg model and points out that the development and utilization of compensation value system should include intergenerational compensation and strategic value besides economic value and ecological value.

Based on the researches above, this paper analyzes market structure of preponderant mental mineral resources such as tungsten, molybdenum, tin, antimony, and rare earth and integrates social preference into Cournot production decision model to analyze the impact exerted on market structure, production decision of developers, price, and profits by social profits, thus discovering the existence of new equilibrium, intergenerational compensation, and strategic value, which clears the agreement pricing mechanism of the metal mineral resources and reveals the pricing power routes affected by intergenerational equity and social preferences.

THE FUNCTION ROUTES OF INTERGENERATIONAL EQUITY TO PREPONDERANT METAL MINERAL RESOURCES PRICING POWER

Cournot Market Structure Analysis of Preponderant Metal Mineral Resources

Cournot, French mathematical economist, first outlined his theory of duopoly market in 1838. In this situation, there exist two enterprises supplying homogeneous products in the market. Each enterprise could choose optimal production to maximize profits by observing others production. He then discovered that a stable equilibrium occurs where each enterprise chooses the production as their rival expected. So the model has a series of strict assumptions: the market is only dominated by two rational suppliers aiming at profit maximization; Oligarch production competition is strategic for supposing each other's output expectation function and price determined by market production; Oligarch determines their own production after prediction and assumes the output of their rival is fixed; the cost of production of oligarchs is zero and marginal cost of production is a certain constant; there is a linear demand function in the market.

Inverse linear demand function in duopoly market is assumed as follows:

$$p(Q) = a - bQ,$$

(1)

Where Q is the total supply of homogeneous products in duopoly market: $Q = q_1 + q_2$ and p is the market price. The output of oligopolist 1 is q_1, the output of oligopolist 2 is q_2, spontaneous demand is a, sensitivity coefficient to price of demand is b. The profits of oligopolists are

$$\pi_i(q_i, q_j) = p(Q) q_i - c_i q_i, \quad i = 1, 2; \; i \neq j.$$

(2)

The marginal cost of production of two oligopolists $c_i > 0$ meets $a > \max(c_1, c_2)$, $b > 0$. If oligopolist has the same marginal cost of production and chooses optimal output independently, then the profits of each oligopolist are

$$\pi_1 = q_1(p - c) = q_1(a - b(q_1 + q_2) - c)$$
$$= -bq_1^2 + (a - c)q_1 - bq_1q_2,$$
$$\pi_2 = q_2(p - c) = q_2(a - b(q_1 + q_2) - c)$$
$$= -bq_2^2 + (a - c)q_2 - bq_1q_2.$$

(3)

Since oligopolists are pursuing profits maximization, first order condition is

$$\frac{\partial \pi_1}{\partial q_1} = -2bq_1 + (a - c) - bq_2 = 0,$$

$$\frac{\partial \pi_1}{\partial q_2} = -2bq_2 + (a - c) - bq_1 = 0.$$

(4)

Combine the above two equations, and we could obtain equilibrium outputs and profits of oligopolists:

$$q_1 = q_2 = \frac{a - c}{3b}, \qquad \pi_1 = \pi_2 = \frac{(a - c)^2}{9b}. \tag{5}$$

Thus we could obtain Cournot equilibrium under the condition of complete information. Equilibrium outputs of oligopolists q_1 and q_2 are optimal output assumed fixed output of their rival, so Cournot equilibrium is a subset of Nash equilibrium.

According to market concentration CR_2 and CR_4 of preponderant metal mineral resources in China, these resources are monopolistic and each oligopolist according to its own profit maximization makes decision simultaneously in oligopoly market. Therefore, this paper uses Cournot game model to analyze the influence exerted on development compensation value and pricing mechanism by combinational equilibrium evaluation factors, to analyze the function routes of combinational equilibrium evaluation factors to preponderant metal mineral resources pricing power and follow the classical assumptions of Cournot equilibrium, that is, assuming oligarchs marginal production cost c is equal. According to the market supply and demand situation and national industrial policy, this paper analyzes the relationship between demand and price of preponderant metal mineral resources by using regression analysis, which shows the feasibility to simulate product demand function by linear demand function in oligopoly market. Therefore, it is assumed that linear inverse demand function of metal mineral resources products is $p = a - q_1 - q_2$, q_1 is the output of oligopolist 1, and q_2 is the output of oligopolist 2 and satisfies the spontaneous demand of the market $a > c$, so oligarchs profits objective functions ignoring psychological preferences of players are as follows:

$$f_1(q_1, q_2) = q_1(a - q_1 - q_2) - cq_1, \tag{6}$$

$$f_2\left(q_1, q_2\right) = q_2\left(a - q_1 - q_2\right) - cq_2,$$

(7)

Where a in (6) and (7) stands for spontaneous demand of metal minerals; function f_1 and f_2 stands for profit function of oligopolists, respectively.

Combine (6) with (7), we could obtain Cournot game equilibrium of each oligopolist:

$$\left(q_1, q_2\right) = \left(\frac{a - c}{3}, \frac{a - c}{3}\right).$$

(8)

Intergenerational Compensation Modification of Preponderant Metal Mineral Resources Development and Compensation

The essence of the preponderant metal mineral resources depletion compensation is value compensation to future losses aiming at excessive mining contemporarily. According to equity theory, externalities of different economic subjects could be solved through negotiations in preponderant metal mineral resources development. If there is reasonable institutional arrangements, the externalities could be internalized to a great extent. However, the externality in mineral resources development for the contemporary is better than the descendant, and because the latter is absent in game negotiation, they could not restrict behavior of the contemporary which cause asymmetry between behaviors. In order to solve the internalization of intergenerational externality problem under the condition of asymmetry, we could build the sustainable development compensation fund in the process of mineral resources development on the basis of the theory of Hotelling mineral resources depletion compensation and Howarth intergenerational property transfer theory. Sustainable development compensation fund is a cash conversion mode; if discount rate is considered, it will keep growing. If intergenerational compensation cost is F, time horizon for compensation is T, and then the intergenerational compensation fund needed is $s = F/(1 + R)^T$ and R is social discount rate.

With the development of world economy, the preponderant metal mineral resources are scarcer, and many countries are looking for a new substitute to get rid of the dependence on metal mineral resources. From the perspective of sustainable development, this research input could affect development routes and improve efficiency of preponderant metal mineral resources, to ensure the rights and interests of future generations. In consequence, research input of substitute should be regarded as part of the intergenerational development compensation value. Research input of substitute contributes to lower current consumption of metal mineral resources from the aspect of metal mineral resources recycling and extends the development and utilization period to meet the needs of metal mineral resources for both the contemporary and the descendent from the aspect of substitute researches.

The Function Routes of Intergenerational Equity Compensation to Preponderant Metal Mineral Resources Strategic Equilibrium Price

Take intergenerational equity value, that is, sustainable development of the compensation fund s ($s>0$) as intergenerational equity compensation. Metal mineral resources development and utilization cost become the combination of marginal production cost and marginal external cost, namely, $c + s$, and inverse demand function of mineral resources products in international market is $p = a - q_1 - q_2$ and satisfies $a > c + s$; the objective function of each country is

$$f_1(q_1, q_2) = q_1(a - q_1 - q_2 - c - s),$$

(9)

$$f_2(q_1, q_2) = q_2(a - q_1 - q_2 - c - s).$$

(10)

Combined (9) with (10), we could obtain Cournot equilibrium from the intergenerational compensation perspective:

$$\left(q_1^*, q_2^*\right) = \left(\frac{a - c - s}{3}, \frac{a - c - s}{3}\right).$$

(11)

Compare (11) to (8); it could be deduced that international trade price of preponderant metal mineral resources should be included into intergenerational compensation modification so as to show its depletion cost of metal minerals. In this way, the supply of preponderant metal mineral resources will decrease, and international trade price will increase. Besides, the greater the intergenerational equity compensation is, the higher the degree of market monopoly will be, so the initial price of metal mineral prices should be higher.

The intergenerational equity compensation of research inputs mainly considers the effects on metal mineral resources development and utilization by technical progress, which would give rise to the appearance of new substitute and affect price elasticity of demand of the replaced metal mineral resources products. Given impacts of substitute, the demand equation of new metal mineral resources products is $pt = a'$

$- q_{1t} - q_{2t}$

The higher the metal mineral resource price is, the more obvious the substitution will be. Then, the trigger point will appear at a rather low price, that is, $a' < a$, reaching the new equilibrium as follows:

$$\left(q_{1t}^*, q_{2t}^*\right) = \left(\frac{a' - c}{3}, \frac{a' - c}{3}\right)$$

(12)

Compare (12) to (8); we could deduce that total market output of metal mineral resource products is smaller when intergenerational compensation cost of substitutes is considered. Besides, the more the research input of substitutes is, the higher the degree of market monopoly will be, so the initial price of metal mineral prices should be higher.

There is no sustainable development compensation fund established to consider intergenerational equity compensation, and no account set up for substitutes research input in accounting system, resulting in underestimation of development compensation costs and deficiency in intrinsic value compensation. In reality, the lower metal premium on mineral resources leads to lower international trade prices. And due to low entry barriers, the development of metal mineral resources exists many problems, such as, small scale, operation chaos and overexploitation, which generate excessive competition and vicious circle to further price reduction.

THE FUNCTION ROUTES OF SOCIAL PREFERENCE TO PREPONDERANT METAL MINERAL RESOURCES PRICING POWER

The Utility Function Modification in Decision Making of Preponderant Metal Mineral Resources Development

Under the imperfect competition market structure, fair price reflects not only intrinsic value compensation equity, industrial organization trade forces equity, and policies trading forces equity in oligopoly market structure of preponderant metal mineral resources, but also supply and demand of intrinsic value compensation equilibrium fluctuations caused by the above equities. From the perspective of behavioral economics, the influences exerted on shadow price and profits by fair belief of stakeholders should be considered in the trade forces equity. As to measurement of strategic value and equity level, it is advisable to learn from the establishment of social utility function. For example, in strategic production decision, if a resource developer has reciprocal preference hopes that the production of its competitor is more than equity output accepted by players, the oligopolist is willing to reduce profits of competitors by squeezing its own profits; if a resource developer has reciprocal preference hopes that the production of its competitor is less than equity output accepted by players, the oligopolist is willing to increase profits of competitors by squeezing its own profits. The profits variation above is the producer surplus variation; hence, it is possible to measure strategic value produced by psychological preferences by the variation of preponderant metal mineral resources developers surplus caused by price variation. Under oligopoly market structure, the modification of psychological preferences to developer decision-making utility function should be under the condition of interdependence preferences; metal mineral resources developer output decision-making utility function included into psychological effects of social preferences is as follows:

$$U_i\left(O\left(s_i, s_i^*\right)\right) = \pi_i\left(O\left(s_i, s_{-i}^*\right)\right)$$

$$+ \sum_{j \neq i} w_{ij}\left(s_i, s_{-i}^*\right) \pi_j\left(O\left(s_i, s_{-i}^*\right)\right) \tag{13}$$

In (13), where $O(s_i, s_i^*)$ is the output decision under interdependence strategy, s_i is the output strategy of oligopolist, i, s_{-i}^* is the output strategy of remaining oligopolists, π_i is oligopolist i's profits without considering interdependent preferences, π_j is the profit of other oligopolists without considering interdependence preferences, and w_{ij} is the coefficient of strategic interaction measuring the profit that oligopolist i gives to other oligopolists. Positive values of the coefficient w_{ij} mean that player j is willing to sacrifice his payoff from outcomes in order to increase the payoff of player j. Negative values mean that player i is willing to sacrifice his payoff from outcomes in order to lower player j's payoff. In addition, $w_{ij}(s_i, s_{-i}^*)\pi_j(O(s_i, s_{-i}^*))$ can be decided by the different types of social preferences as follows.

- If the oligopolist prefers altruism fairness, that is, the oligopolist considers the intertemporal allocation of preponderant mental mineral resources development and the utilization of later generations, then the oligopolists have slight altruistic preferences, and w_{ij} is positive.

- For types of inequity averse player, $w_{ij}(s_i,s^*_{-i})\pi_j(O(s_i,s^*_{-i}))$ can be replaced by , $w_{ij}(q_i,Q^*_{-i})(\pi_j - \pi_i), w_{ij}(q_i,Q^*_{-i})$ is used to measure the deviation profit function of oligopolist i puting weights on oligopolist j, and here is

$$w_{ij}\left(q_i,Q^*_{-i}\right)\begin{cases} > 0, & \pi_j < \pi_i \\ = 0, & \pi_j = \pi_i \\ < 0, & \pi_j > \pi_i. \end{cases}$$

(14)

The first condition expresses aversion to advantageous inequity, namely, if oligopolist i's profits are greater than those of oligopolist j, then oligopolist i is willing to sacrifice its own profits to increase j's profits. The third condition expresses aversion to disadvantageous inequity. If oligopolist i's profits are lower than those of oligopolist j, then oligopolist i is willing to sacrifice its own profits to reduce j's profits.

- If it is the reciprocal fairness preference, the payoff function of the oligopolist i is $U_i(q_i,Q_{-i}) = \pi_i(q_i,Q_{-i}) + w_i(Q_{-i},Q^F_{-i})\sum_{j\neq i}\pi_i(q_i,Q_{-i})$. Where $\pi_i(q_i,Q_{-i})$ is oligopolist i's profits and is the weight that oligopolist i places on its rivals gross profits, that is, $\sum_{j\neq i}\pi_i(q_i,Q_{-i})$, and on the gross output of its rivals Q_{-i}, the equation is $\pi_i(q_i,Q_{-i}) = R_i(q_i,Q_{-i}) - C_i(q_i),,$ where $R_i(q_i,Q_{-i}) = P(Q)q_i$ is revenue. Assuming that oligopolist i is endowed with the weight on its rivals depending on fair gross output Q^F_{-i} and output of his rivals. Furthermore, it can be assumed that

$$w_i\left(Q_{-i},Q^F_{-i}\right)\begin{cases} > 0, & Q_{-i} < Q^F_{-i} \\ = 0, & Q_{-i} = Q^F_{-i} \\ < 0, & Q_{-i} > Q^F_{-i}. \end{cases}$$

(15)

That is, when $Q_{-i} < Q_{-i}^F$ the oligopolist i has a positive weight on rivals' gross profits; when $Q_{-i} = Q_{-i}^F$, the weight is 0; and it has a negative weight on its rivals' output when $Q_{-i} > Q_{-i}^F$. These conditions reveal the real intention of oligopolist with reciprocal fairness preference to care rivals. The Rabin fairness equilibrium determination method used by reciprocal fairness psychological compensation value modification is that game subjects are willing to sacrifice their material interests to help people who treat them kindly and to punish people who treat them badly; the smaller the sacrifice, the greater motivation to help and punish.

The Function Routes of Social Preference to Preponderant Metal Mineral Resources Strategic Equilibrium Price

Based on the revised developers' utility function, the developers will play strategic reciprocal game on production when exploiting preponderant mental mineral resources; meanwhile they can tell the industry is oligopoly by judging from the market concentration indicators of CR_2 and CR_4 of preponderant metal mineral resources. Thus, developers will play oligopolistic reciprocal fairness game, and each oligopolist based on profit maximization principle to make decision simultaneously. Therefore, the Cournot game model is fit to analyze the function routes of social preferences to preponderant metal mineral resources pricing power improvement.

The Function Routes of Altruism Preference to Preponderant Metal Mineral Resources Strategic Equilibrium Price

Preponderant metal mineral resources development requires sustainable development, so it could be assumed that slight altruistic preference is possessed on the consideration of intergenerational equity. According to (13) and Cournot hypothesis, monopoly profit functions under altruism preference of preponderant metal mineral resources development are

$$\pi_1 \left(q_1, q_2, \lambda_1\right) = q_1 \left(a - q_1 - q_2 - c - s\right)$$
$$+ \lambda_1 q_2 \left(a - q_1 - q_2 - c - s\right), \qquad (16)$$

$$\pi_2 \left(q_1, q_2, \lambda_2\right) = q_2 \left(a - q_1 - q_2 - c - s\right)$$
$$+ \lambda_2 q_1 \left(a - q_1 - q_2 - c - s\right), \qquad (17)$$

Where $\lambda_i (i = 1, 2)$ is oligopolist i's slight altruism preference coefficient, and $\lambda_i \geq 0$. Altruism preference coefficients fall in the interval (0, 1) approximately revealed by game experiments according behavior of experimental economics and psychology, such as trust game, gift exchange game, dictator game, and market game with punishment or without punishment. According to the optimal Cournot equilibrium analysis method, the optimal reaction function of each oligopolist could be gained from (16) and (17):

$$\frac{\partial \pi_1}{q_1} = \left(a - 2q_1 - q_2 - c - s\right) - \lambda_1 q_2 = 0, \qquad (18)$$

$$\frac{\partial \pi_2}{q_2} = \left(a - 2q_2 - q_1 - c - s\right) - \lambda_2 q_1 = 0. \qquad (19)$$

Combine (18) with (19), we could obtain Cournot equilibrium under pure altruism preference:

$$\left(q_1^{**}, q_2^{**}\right)$$

$$= \left(\frac{(1-\lambda_1)(a-c-s)}{4-(1-\lambda_1)(1-\lambda_2)}, \frac{(1-\lambda_2)(a-c-s)}{4-(1-\lambda_1)(1-\lambda_2)} \right). \tag{20}$$

Using Cournot equilibrium output under altruism preference of oligopolist 1 minus that in (11), we could get the equation:

$$\frac{(1+\lambda_1)(a-c-s)}{4-(1-\lambda_1)(1-\lambda_2)} - \frac{(a-c-s)}{3}$$

$$= \frac{(a-c-s)(-2\lambda_1-\lambda_2+\lambda_1\lambda_2)}{3\left[4-(1-\lambda_1)(1-\lambda_2)\right]}. \tag{21}$$

Since $(\lambda_1, \lambda_2) \to 0$, $(a-c-s)/3[4-(1-\lambda_1)(1-\lambda_2)] > 0$, and $(-2\lambda_1-\lambda_2+\lambda_1\lambda_2) < 0$, the Cournot equilibrium output under pure altruism condition is lower than that under no altruism condition. And the higher the degree of altruism preference is, the smaller the total market output will be, so strategic price of preponderant metal mineral resources development and compensation should be higher.

The Function Routes of Inequity Aversion to Preponderant Metal Mineral Resources Strategic Equilibrium Price

In the development compensation pricing of preponderant mental mineral resources, oligopolists show sympathy preference and jealousy preference on the players' payoffs, say, they will sacrifice their profits to lower those oligopolists who obtain higher profits, but also sacrifice their profits to upgrade those oligopolists who bear lower profits. According to Fehr and Schmidt's definition of inequity aversion, the payoff functions of preponderant mental mineral resources development and utilization are affected by oligopolists' inequity aversion preferences, and thus their payoff functions are

$$\pi_1 (q_1, q_2, \alpha_1, \beta_1) = f_1 - [a_1 \max (f_2 - f_1, 0)$$

$$+ \beta_1 \max (f_1 - f_2, 0)],$$

$$\pi_2 (q_1, q_2, \alpha_2, \beta_2) = f_2 - [a_2 \max (f_1 - f_2, 0)$$

$$+ \beta_1 \max (f_2 - f_1, 0)],\tag{22}$$

Where α_i $(i = 1, 2)$ is the jealousy preference coefficient under inequity aversion of oligopolist i and βi $(i = 1, 2)$ is the sympathy preference coefficient under inequity aversion; moreover, $\alpha_i > \beta_i > 0$. And owing to the symmetry form in oligopoly market structure, the assumption of $f_2 > f_1$ will not affect analysis conclusion; thus (22) are as follows:

$$\pi_1 (q_1, q_2, \alpha_1) = q_1 (a - q_1 - q_2 - c - s)$$

$$- \alpha_1 [q_2 (a - q_1 - q_2 - c - s)$$

$$- q_1 (a - q_1 - q_2 - c - s)]$$

$$\pi_2 (q_1, q_2, \alpha_2) = q_2 (a - q_1 - q_2 - c - s)$$

$$- \beta_2 [q_2 (a - q_1 - q_2 - c - s)$$

$$- q_1 (a - q_1 - q_2 - c - s)],\tag{23}$$

Where α_i $(i = 1, 2)$ is the jealousy preference coefficient under inequity aversion of oligopolist i and β_i $(i = 1, 2)$ is the sympathy preference coefficient under inequity aversion. The second items on the right side in (23) are disutility produced by oligopolist i's jealousy preference. According to the optimal Cournot equilibrium analysis, the optimal response function of each oligopolist derived from revenue functions under inequity aversion is as follows:

$$\frac{\partial \pi_1}{q_1} = (a - 2q_1 - c - s)(1 + \alpha_1) - q_2 = 0,\tag{24}$$

$$\frac{\partial \pi_2}{q_2} = (a - 2q_2 - c - s)(1 - \beta_2) - q_1 = 0.$$

(25)

Combine (24) with (25); the optimal production of each oligopolist preferring inequity aversion can be finally written as

$$\left(q_1^{***}, q_2^{***}\right) = \left(\frac{(1 + 2\alpha_1)(1 - \beta_2)(a - c - s)}{4(1 + \alpha_1)(1 - \beta_2) - 1}, \right.$$

$$\left. \frac{(1 + \alpha_1)(1 - 2\beta_2)(a - c - s)}{4(1 + \alpha_1)(1 - \beta_2) - 1}\right).$$

(26)

The fairness equilibrium output function exhibits that under piecewise linear inequity aversion condition, the optimal response function of oligopolist and standard Cournot equilibrium game are both continuous, but the former is no longer monotonous.

Using Cournot equilibrium output in (26) minus that in (11), we could get the equation:

$$q_1^{***} - q_1^{**} = \frac{2a_1 + \beta_2 - 2a_1\beta_2}{12(1 + \alpha_1)(1 - \beta_2) - 3}(a - c - s),$$

(27)

$$q_2^{***} - q_2^{**} = \frac{-a_1 - 2\beta_2 - 2a_1\beta_2}{12(1 + \alpha_1)(1 - \beta_2) - 3}(a - c - s).$$

(28)

The result of (28) is obviously less than zero. After thousands of game experiments, such as ultimatum game, dictator game, and public good games in different countries, it proves that about 85% of the people's α_1 and β_1 fall in the interval (0.15, 0.50), so the result of (27) is more than zero. Equation (27) shows that Cournot equilibrium output preferring fairness is more than that when they are only concerned about their own enterprise profits, while (28) shows that Cournot equilibrium output preferring fairness is less than that when they are only concerned about their own enterprise profits. Such results indicate the effects of sympathy and jealousy preference on the Cournot equilibrium, and weak complementary between degree of oligopolist's sympathy and equilibrium output, that is, when jealousy preferences is greater, the optimal Nash equilibrium market output under inequity aversion in Cournot game will be larger. Under the circumstances, the improvement of jealousy preferences reduces the producer surplus and increases consumer surplus. On the other side, the minimal Nash equilibrium market output under piecewise linear inequity aversion in Cournot game will decrease when sympathy preferences are greater. In this case, the improvement of sympathy preferences increases the producer surplus and reduces consumer surplus. The variation of producer surplus is greater than that of consumer surplus after considering fairness preference, thus producing strategic reciprocal value for oligopolists. Therefore, it is necessary to take compensation of strategic reciprocal value in price system into account in the pricing process of preponderant metal mineral resources.

The Function Routes of Reciprocal Equity Equilibrium to Preponderant Metal Mineral Resources Strategic Equilibrium Price

Under the condition of intergenerational equity equilibrium and according to the definition of reciprocal equity equilibrium, the players' revenue function of preponderant metal mineral resources development is

$$\pi_1\left(q_1, q_2\right) = q_1\left(a - q_1 - q_2 - c - s\right);$$

$$\pi_2\left(q_1, q_2\right) = q_2\left(a - q_1 - q_2 - c - s\right)$$

(29)

$$\pi_1^h(q_1) = q_1(a - q_1 - c - s);$$

$$\pi_1^e(q_1) = \frac{q_1(a - q_1 - c - s)}{2}$$

$$(30)$$

$$\pi_2^h(q_2) = q_2(a - q_2 - c - s);$$

$$\pi_1^e(q_2) = \frac{q_2(a - q_2 - c - s)}{2},$$

$$(31)$$

where, $\pi_1^l(q_1) = 0, \pi_1^{min}(q_1) = 0, \pi_2^{min}(q_2) = 0$. According to the definition of reciprocal equity equilibrium, in the transaction of preponderant metal mineral resources, the friendliness function between oligopolist 1 and oligopolist 2 is

$$f_1(q_1, q_2) = \frac{1}{2} - \frac{q_1}{a - q_2 - c - s},$$

$$f_2(q_1, q_2) = \frac{1}{2} - \frac{q_2}{a - q_1 - c - s}.$$

$$(32)$$

Friendliness belief between oligopolist 1 and oligopolist 2 is

$$\tilde{f}_2(\tilde{q}_1, q_2) = \frac{1}{2} - \frac{q_2}{a - \tilde{q}_1 - c - s},$$

$$\tilde{f}_1(q_1, \tilde{q}_2) = \frac{1}{2} - \frac{q_1}{a - \tilde{q}_2 - c - s}.$$

$$(33)$$

According to the definition of (29) and (33), the utility functions of different players in preponderant metal mineral resources development are

$$U_1\left(q_1, q_2, \tilde{q}_1\right) = \pi_1\left(q_1, q_2\right) + \tilde{f}_2\left(\tilde{q}_1, q_2\right) + \left[1 + f_1\left(q_1, q_2\right)\right]$$

$$= q_1\left(a - q_1 - q_2 - c - s\right)$$

$$+ \left[\frac{1}{2} - \frac{q_2}{a - \tilde{q}_1 - c - s}\right]$$

$$\times \left[\frac{3}{2} - \frac{q_1}{a - q_2 - c - s}\right], \tag{34}$$

$$U_1\left(q_1, q_2, \tilde{q}_1\right) = \pi_2\left(q_1, q_2\right) + \tilde{f}_1\left(q_1, \tilde{q}_2\right)\left[1 + f_1\left(q_1, q_2\right)\right]$$

$$= q_2\left(a - q_1 - q_2 - c - s\right)$$

$$+ \left[\frac{1}{2} - \frac{q_1}{a - \tilde{q}_2 - c - s}\right]$$

$$\times \left[\frac{3}{2} - \frac{q_2}{a - q_1 - c - s}\right]. \tag{35}$$

Combine (34) with (35) to get the first-order optimal solution, thus obtaining Cournot equilibrium solution under reciprocal equity equilibrium:

$$q_1^{****} = \frac{1}{2} \left(\frac{4a - 4c - 4s}{3} - \frac{\sqrt{3 + (a - c - s)^2}}{3} \right.$$

$$\left. - \frac{2(ac + as)\sqrt{3 + (a - c - s)^2}}{9} \right)$$

$$+ \frac{1}{2} \left(\frac{(c + s)^2 \sqrt{3 + (a - c - s)^2}}{9} \right.$$

$$+ \frac{a^2 \sqrt{3 + (a - c - s)^2}}{9}$$

$$\left. - \frac{\left[3 + (a - c - s)^2 \right]^{3/2}}{9} \right),$$

$$q_2^{****} = \frac{1}{3} \left(2a - 2c - 2s - \sqrt{3 + (a - c - s)^2} \right).$$

(36)

It could be seen in (30) that $q_2^{****} < (1/3)(2a - 2c - 2s - \sqrt{(a - c - s)^2}) = (a - c - s)/3 = q_2^{**}$, similarly, $q_1^{****} < q_1^{**}$. It suggests that after considering reciprocal equity preferences, the total market output of preponderant metal mineral resources is reduced and the degree of market monopoly is higher. The demand price elasticity of preponderant metal mineral resources is rather small for the reason that they are industrial raw materials and hard to be replaced. Given reciprocal equity equilibrium, the market capacity is decreased and then caused higher prices. The greater producer surplus is strategic reciprocal value produced by reciprocal equity preference, which should be included in compensation value system. If oligopolist 1 and oligopolist 2 are present two countries, it is observed that two countries having reciprocal equity intention could enhance their metal mineral resources market monopoly status, thereby obtaining pricing power for themselves. It also explains the reason why China should obey reciprocal equality principles in international trade.

SIMULATION ANALYSIS

As has been observed from (20), (26), and (36), oligarchs' social preferences in preponderant metal market, that is, altruism preference, inequity aversion, and reciprocal fairness, produce psychological effects, which raise the power of oligarch market and change output decision (reducing production) of each oligarch by being blend in decision function. Therefore, market capacity is decreased and price of supply and demand is increased. Considering the significance of preponderant metals, their demand is rigid, so developers has larger producer surplus, which is strategic reciprocal value produced by reciprocal fairness. However, it can be seen from equilibrium results of the above four equations that forms of strategic value are various. So the method of numerical simulation is used to verify as follows.

Original Basic Parameter Setting

According to the market supply and demand of lithium, antimony, indium, rare earth, and the national industrial policy as well as the regression analysis of the Cournot linear demand function, the spontaneous demand stabilized at around 2,000 tons, so a in the Cournot model can take the value 2000, namely, a=2000, by analyzing the tax subjects of preponderant mental mineral resources development, namely, the property cost (mineral resources compensation, resource tax), mining costs (outlay of exploration, outlay of mining), investment capital (capital investment per ton of mineral resources), production costs (raw materials, power costs, wages and benefits, manufacturing costs, processing fees, finance charges, and operating expenses), security costs (safety training, disinfection equipment, risk assessment costs, occupational funds, and pension), and part of the measurable environmental governance operating costs (water pollution, air pollution, waste pollution, and heavy metal pollution), and environmental restoration costs (mine land reclamation bond, tailings management costs, and mine environmental geology warning inputs). Based on tax subjects above and the statistical analysis of preponderant metal development compensation enterprises, the basic cost c of preponderant mental mineral resources development compensation enterprises is about 800 units; original basic parameter setting in Section

4.1 is obtained by regression estimation in last 10 years, so it conforms to the market situation. Thus the results of numerical simulation could support the application of practical program. The sustainable development fund is used to measure modification of intergenerational compensation, estimates depletion costs of rare earth, lithium, and indium by modified user cost approach and gains the proportion of depletion cost to total cost is 20%. The depletion cost is the reflection of intergenerational compensation modification actually, so the value of sustainable development fund s of intergeneration equity is 160 units.

Impacts Exerted on Cournot Equilibrium by Coefficients of Altruism Preference and Inequity Aversion

Reciprocal equity Cournot equilibrium variation under reciprocal equity preference has the consistent results with classic Cournot equilibrium, so numerical simulation is not needed to analyze its character. However, Cournot equilibrium under social preferences produces a new equilibrium bringing about a new character, on account of the variation of altruism preferences and the loss aversion of inequity aversion; thus, numerical simulation is required. Altruism preference coefficients fall in the interval (0, 1) revealed by game experiments and modified game experiments of altruism preference and inequity aversion preference, such as, ultimatum game, dictator game, public good games, gift exchange game, and third-party punishment game. While the distribution of inequity aversion coefficients has the following features, as is shown in Table 1.

Table 1: The distribution of jealousy and sympathy preferences coefficients α and β under inequity aversion

Value and proportion of α		Value and proportion of β	
Value of α	Proportion (%)	Value of β	Proportion (%)
$\alpha=0$	20	$\beta=0$	30
$\alpha=0.5$	65	$\beta=0.25$	45
$\alpha=1$	10	$\beta=0.6$	10

α=4	5	β=0.6	5

Simulation Results of Impact Exerted on Cournot Equilibrium by Altruism Preference

Altruism preference coefficients could be obtained by game experiments and impacts exerted on Cournot equilibrium decision-making by altruism preference could be simulated by (20) and (21). The specific content is shown in Figure 1.

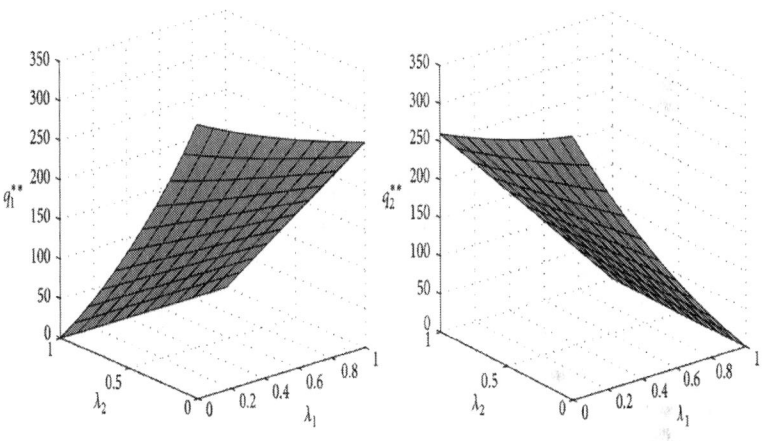

Figure 1: The sensitivity analysis of Cournot equilibrium output to altruism preference coefficient.

As can be seen from Figure 1, the Cournot equilibrium output under altruism preference modified by intergenerational equity is changing as follows. In a certain condition, the altruism degree of oligopolist 2 is in proportion to that of oligopolist 1, namely, the bigger the altruism coefficient of oligopolist 2 is, the smaller the Cournot equilibrium output of oligopolist 1 is, and vice versa. This result is in accordance with the experiment result of pure altruism preference. According to Cournot equilibrium and analysis framework of supply and demand, the reduction of Cournot equilibrium output will raise price in preponderant mental market, while to those preponderant mental resources lacking price elasticity of demand, price increase will lead to producer surplus increase, which is the strategic value of development and compensation of preponderant metal resources.

Simulation Results of Impact Exerted On Cournot Equilibrium by Inequity Aversion

Inequity aversion coefficients could be obtained by game experiments and impacts exerted on Cournot equilibrium decision-making by inequity aversion which could be simulated by (26) and (28). The specific content is shown in Figure 2.

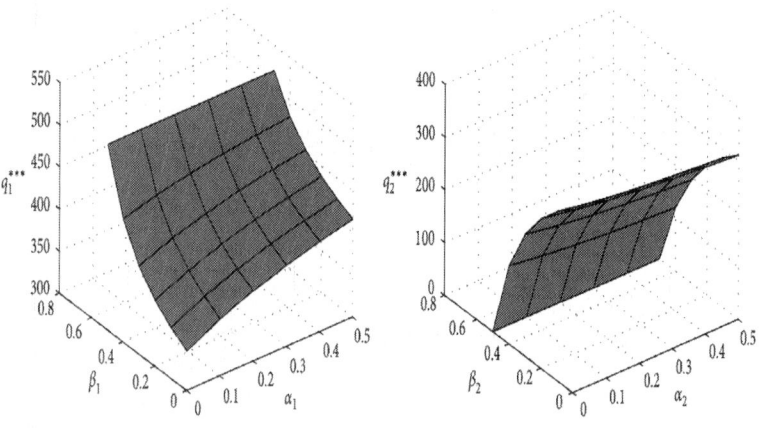

Figure 2: The sensitivity analysis of Cournot equilibrium output to inequity aversion preference.

As can be seen from Figure 2, Cournot equilibrium output under intergenerational equity correction when inequity aversion is considered is smaller than that when they are not considered. The oligopolists under inequity aversion lift the degree of market monopoly of preponderant metal mineral resources. Besides, the higher the degree of jealousy preference is, the larger the Cournot equilibrium output will be, while as to sympathy preference, the result is completely opposite. These results are consistent with the results of theoretical model; namely, Cournot equilibrium under inequity aversion has the character of loss aversion.

CONCLUSIONS

This paper analyzes the modification of intergenerational equity and social preferences to fiducial value of preponderant metal mineral resources and qualifies the impacts exerted on Cournot equilibrium by interdependence preferences (altruism preference, inequity aversion, and reciprocal equity preference), thus drawing the following conclusions.

- Considering the intergenerational equity and social preferences in preponderant metal mineral resources development, Cournot equilibrium market capacity becomes smaller. Because preponderant metal mineral resources are significant to national security and industrial development and hard to replace, the demand of those prices is rigid. Therefore, when Cournot equilibrium decreases, oligopolist's profits increase; that is, oligopolist's producer surplus increases. The variation of producer surplus is intergenerational equity compensation value and strategic value when intergenerational equity and social preferences blend into Cournot game.

- However, the impact exerted on Cournot market power by social preferences is inconsistent. Variation of altruism Cournot equilibrium and reciprocal equity Cournot equilibrium are consistent, while Cournot equilibrium under inequity aversion has the characteristic of loss aversion, namely, under the consideration of inequity aversion Cournot competition, Cournot Nash equilibrium transform monotonically with sympathy and jealousy inequity aversion; that is, if the jealousy degree of oligopoly increased, Cournot Nash equilibrium is near the perfect competition output equilibrium; if the sympathy degree of oligopoly increased, Cournot Nash equilibrium is near the optimal collusion equilibrium output.

The results indicate that the essence of price distortion of preponderant metal mineral resources is incomplete value realization and resource value compensation inequity, failing to realize the goal of mineral resources price reform, namely, two basic conditions for the reform are not satisfied. The production value of mineral resources is achieved through spontaneous effect of the market for its relevance to efficiency. However, the property value, intragenerational value, and intergenerational value of mineral resources are difficult to realize spontaneously on account of their public characteristic under market effect. It is difficult for market-oriented reform of mineral resources price to fully realize the mineral resources value and provide a fundamental guarantee for sufficient and reasonable compensation. Therefore, to achieve the goal of mineral resources price reform, it is necessary to reconstruct value compensation system of metal mineral resources development. According to redefinition of preponderant metal mineral resources development under the principle of multiple equilibrium value evaluation, the actual negotiated pricing mechanism is classified based on mineral resources pricing mechanism of multiple equilibrium evaluation models; that is, pricing should fully reflect the complete elements of mineral resources and coordinate interests between players. As to practical operation scheme of current resources tax reform, value measurement of preponderant metal mineral resources development compensation should analyze from not only development and utilization results but also the perspective of strategic reciprocal psychology. Besides, the value system of preponderant metal mineral resources development compensation contains economic value and ecological value as well as strategic value. Furthermore, since social preferences are added into the value system of preponderant metal mineral resources development compensation, the market monopoly degree will be strengthened and development compensation price will be higher, which require perfect tax subjects and establish full cost theory system, thus estimating mineral resources value correctly and making the international trade fair prices tend to rationalization.

ACKNOWLEDGMENTS

This work was partially supported by the National Natural Science Foundation of China (71171203, 71073177), the major Project of National Social Science Foundation (13&ZD024, 13&ZD169), the Research Project in Humanities and Social Sciences of Chinese Ministry of Education (13YJAZH149, 09YJC90261), the Ph.D. Programs Foundation of Ministry of Education of China (20130162110076), the Emergency Project of Chinese Ministry of Education (2009 JYJR035), the China Postdoctoral Science Foundation (20110491264), the Special Project of Postdoctoral Research Funds of Hunan Province

(S2011R1040), and the Project of Soft Science of Hunan Province (2009ZK3193).

REFERENCES

1.   J. T. Hu, "Hold high the great banner of socialism with Chinese characteristics and strive for new victories in building a moderately prosperous society in all respects," in Proceedings of the 17th National Congress of the Communist Party of China, Beijing, China, 2007.

2.   J. T. Hu, "Firmly march on the path of socialism with Chinese characteristics and strive to complete the building of a moderately prosperous society in all respects," in Proceedings of the 18th National Congress of the Communist Party of China, Beijing, China, 2012.

3.   H. Hotelling, "The economics of exhaustible resources," Journal of Political Economy, vol. 39, no. 2, pp. 137–175, 1931.

4.   J. M. Hartwick, "Intergenerational equity and the investing of rents from exhaustible resources," The American Economic Review, vol. 67, no. 5, pp. 972–974, 1977.

5.   E. Serafy, "Absorptive capacity, the demand for revenue, and the supply of petroleum," The Journal of Energy and Development, vol. 7, no. 1, pp. 73–88, 1981.

6.   E. Serafy, "The proper calculation of income from depletable natural resources," in Environmental Accouting for Sustainable Development, a UNEP-World Bank Sympoisum, Y. J. Ahmed, S. E. Seral, and E. Lutz, Eds., The World Bank, Washington, DC, USA, 1989.

7.   M. A. Adelman, "User cost in oil production," Resources and Energy, vol. 13, no. 3, pp. 217–240, 1991.

8.   C. E. F. Young and R. Seroa Da Motta, "Measuring sustainable income from mineral extraction in Brazil," Resources Policy, vol. 21, no. 2, pp. 113–125, 1995.

9.   J. N. Blignaut and R. M. Hassan, "Assessment of the performance and sustainability of mining sub-soil assets for economic development in South Africa," Ecological Economics, vol. 40, no. 1, pp. 89–101, 2002.

10. B. Q. Lin, X. Y. Liu, C. Y. Zou, and X. Liu, "Resource tax reform: a case study of coal from the perspective of resource economics," Social Sciences in China, vol. 18, no. 3, pp. 116–139, 2012.

11. G. P. Li and H. W. Li, "The perfection of user cost approach and the estimation of user cost of oil and gas resources of America," Journal of Natural Resources, vol. 28, no. 6, pp. 1046–1058, 2013.

12. X. F. Zeng and G. P. Li, "A new estimation of user costs for non-renewable energy resources," Resources Science, vol. 35, no. 2, pp. 439–446, 2013.

13. B. Fattouh, OPEC Pricing Power: The Need for a New Perspective, vol. 158, Oxford University Press, 2007.

14. R. K. Kaufmann, "The mechanisms for autonomous energy efficiency increases: a cointegration analysis of the US energy/GDP ratio," The Energy Journal, vol. 25, no. 1, pp. 63–86, 2004.

15. A. Rubinstein, "Perfect equilibrium in a bargaining model," Econometrica, vol. 50, no. 1, pp. 97–109, 1982.

16. F. Wen, X. Gong, Y. Chao, and X. Chen, "The effects of prior outcomes on risky choice evidence from the stock market," Mathematical Problems in Engineering, vol. 2014, Article ID 272518, 8 pages, 2014.·

17. F. Wen and X. Yang, "Skewness of return distribution and coefficient of risk premium," Journal of Systems Science & Complexity, vol. 22, no. 3, pp. 360–371, 2009.

18. F. Wen, Z. He, and X. Chen, "Investors' risk preference characteristics and conditional skewness,"Mathematical Problems in Engineering, vol. 2014, Article ID 814965, 14 pages, 2014. ·

19. F. Wen, Z. He, X. Gong, and A. Liu, "Investors' risk preference characteristics based on different reference point," Discrete Dynamics in Nature and Society, vol. 2014, Article ID 158386, 9 pages, 2014.·

20. J. Zhang, "The problem of rare earth products export pricing power loss and reason analysis," Northern Economy, no. 7, pp. 66–67, 2010.

21. Z. M. Wang and X. J. Zhang, "Research into the effect of rare earth resource tax on the export market power of "oligarch" country," Economic Survey, no. 2, pp. 52–55, 2012.

22. C. F. Wu and S. Jiang, "The efficiency of china's futures market: a study based on overreaction and domestic and international markets linkage," Journal of Financial Research, vol. 2, pp. 49–62, 2007.

23. M. Rabin, "Incorporating fairness into game theory and economics," The American Economic Review, vol. 83, no. 5, pp. 1281–1302, 1993.

24. M. Dufwenberg and G. Kirchsteiger, "A theory of sequential reciprocity," Discussion Paper, Tilburg University, 1998.

25. E. Fehr and K. M. Schmidt, "A theory of fairness, competition, and cooperation," Quarterly Journal of Economics, vol. 114, no. 3, pp. 817–868, 1999.

26. G. E. Bolton and A. Ockenfels, "ERC: a theory of equity, reciprocity, and competition," The American Economic Review, vol. 90, no. 1, pp. 166–193, 2000.

27. M. R. Zhong, J. Y. Chen, X.-h. Zhu, and J.-b. Huang, "Strategic equilibrium price analysis and numerical simulation of preponderant high-tech metal mineral resources," Transactions of Nonferrous Metals Society of China, vol. 23, no. 10, pp. 3153–3160, 2013.

Evaluation of Groundwater Recharge Estimates in a Partially Metamorphosed Sedimentary Basin in a Tropical Environment: Application of Natural Tracers

Felix Oteng Mensah[1], Clement Alo[1,] and Sandow Mark Yidana[2]

[1]Department of Earth and Environmental Studies, Montclair State University, Montclair, NJ 07043, USA

[2]Department of Earth Science, University of Ghana, Legon, Accra, Ghana

ABSTRACT

This study tests the representativeness of groundwater recharge estimates through the chloride mass balance (CMB) method in a tropical environment. The representativeness of recharge estimates using this methodology is tested using evaporation estimates from isotope data, the general spatial distribution of the potential field, and the topographical variations in the area. This study suggests that annual groundwater recharge rates in the area ranges between 0.9% and 21% of annual precipitation. These estimates are consistent with evaporation rates computed from stable isotope data of groundwater and surface water in the Voltaian Basin. Moreover, estimates of groundwater recharge through numerical model calibration in other parts of the terrain appear to be consistent with the current data in this study. A spatial distribution of groundwater recharge in the area based on the estimated data takes a pattern akin to the spatial pattern of distribution of the hydraulic head, the local topography, and geology of the terrain. This suggests that the estimates at least qualitatively predicts the local recharge and discharge locations in the terrain.

INTRODUCTION

In regional hydrogeological studies and groundwater resources assessments, accurate estimates of groundwater recharge are required to ensure proper water balance studies and evaluation of groundwater resources for productive uses. Recharge is one of the uncertain parameters in model calibration and is regarded as one of the major parameters which determine the accuracy and reliability of predictive groundwater flow models. It is therefore apposite that this component of basin wide regional hydrogeological investigations is constrained through a variety of methods. Several methods have been proposed in the literature for providing reliable estimates of groundwater recharge at both the regional and local scales. They range from direct measurements through mass balance techniques and Darcian method, to the use of tracers [1, 2]. A wide variety of mass balance techniques are available and their utilities are often based on an assessment of the validities of underlying assumptions in the areas where such applications are sought. Rorabaugh and Meyboom [3] developed methodologies that are based on baseflow recession and assume amongst others that the baseflow component of stream flow can be approximated to groundwater recharge in large basins, especially where such a hydraulic connection can be established between stream flow and the underlying aquifers. Using this methodology, there is a high possibility of overestimating groundwater recharge rates especially where the catchment area of the stream is larger than the domain being examined. The water table fluctuations method [4] has also been advanced and used in several areas to estimate groundwater recharge in unconfined aquifers. The method is based on the assumption that appreciation in groundwater levels between seasons is attributed to groundwater recharge, which is proportional to the specific yield of the material. Therefore, where the aquifer is confined or semiconfined, the methodology is less appropriate and will lead to underestimation or overestimation depending on the values of the specific yield used. Even where the aquifer is unconfined, the spatial variability in the specific yield field can lead to significant departures of recharge estimates from the reality on the ground. The method is often used in places where initial estimates are required for much more detailed regional hydrogeological studies. Several tracer techniques have been proposed in the literature and they range from the application of isotope techniques to the use of natural hydrochemical conservative tracers such as the chloride ion [2]. The use of tracers to explain groundwater recharge and flow processes is copiously documented in the literature [5]. The chloride mass balance (CMB) technique is based on the assumption that the chloride ion behaves conservatively and is not easily affected by reactions through the unsaturated zone through to the saturated zone. If this assumption is valid, then it follows that the ion can adequately trace groundwater recharge processes and can thus provide reasonable estimates of groundwater recharge in the area. Its reliability therefore hinges on the compatibility of the precipitation events that recharged the system under study and recent precipitation. It is also assumed that the main source of chloride in groundwater is precipitation. Therefore, where it can be determined that a substantial proportion of the groundwater chloride is generated from mineral dissolution processes, the method can lead to underestimation of recharge. On the other hand, where the groundwater chloride is negatively affected by a process which reduces its content compared to that of precipitation, groundwater recharge can potentially be underestimated. The CMB methodology has been widely tested and regarded as one of the most reliable techniques for estimating groundwater recharge in regional hydrogeological studies and basin wide groundwater resources assessments (e.g., [5–10]).

The current study evaluates the performance of the CMB methodology in a typical tropical climatic environment where the availability of groundwater resources is critical to socioeconomic conditions of populations and the survival of ecosystems that depend on such groundwater resources for sustenance. Groundwater recharge estimates from the CMB method in this study are checked against estimates of evaporation rates of percolating rainwater as estimated from isotope techniques. The spatial pattern of variation in the resulting recharge estimates is then predicted using a spatial prediction method for the resource to be evaluated. The novelty lies in the fact that the CMB method, although not new, is rigorously evaluated using another conservative tracer.

THE STUDY AREA

The study area (Figure 1) is in the Northern Region of Ghana and covers a total land area of about 1790.7 Km², and it falls within the White Volta Basin. The agricultural sector is the largest employer in the area, employing about 97% of the active populations [11]. Most of the farmers depend on rain for food production and the typical crops grown include yam, maize, rice, groundnut, cowpea, and soya beans. Unfortunately, agricultural activities are constrained by unreliable rainfall, inadequate irrigation facilities, and difficulty in loan accessibility and lack of storage or processing units leading to postharvest losses [12]. The area is generally flat with gently undulating reliefs. There are no high mountains in the area except few hilly features observed in the southern part with elevations generally ranging between 122 and 244 m above sea level; however, the north is low lying. The area is mainly drained by the White Volta and its tributaries. It falls within the Sudan Savannah zone with an annual rainfall between 900 and 1,200 mm distributed on average over 74 rainy days [13]. The study area falls under the Intertropical Convergent Zone, the interface where two air masses, tropical continental and tropical maritime, overlap. The frontal activity and relative movement of the two air masses control the amount and duration of rainfall. Rainfall, which is generally of short duration and high intensity and is often preceded by thunderstorms and line squalls starts intermittently between March and April through August to September when it turns stable and very heavy. The dry seasons are pronounced with temperature ranging between 18 and 42°C with mean value of approximately 30°C [14]. Relative humidity during wet/rainy season is in the range 40 and 70% and drops to about 15% during the rest of the year [13]. The observed low humidities and high temperatures have led to high potential and actual evapotranspiration rates in the basin.

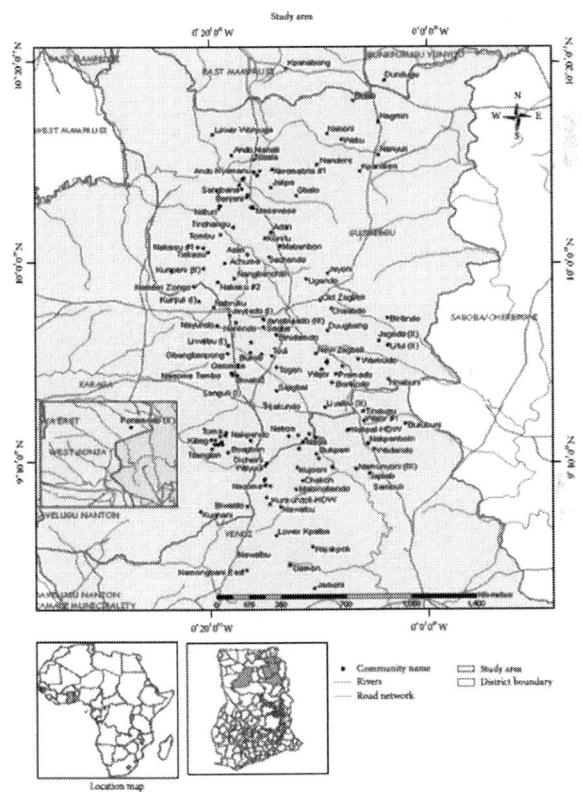

Figure 1: A map of the study area showing the major communities and settlements.

The area is underlain by consolidated Neoproterozoic sedimentary sequences of the Voltaian Super-Group (Figure 2). The rocks are mainly of the Middle Voltaian (Figure 3), which is the most extensive sedimentary formation in Ghana. The Middle Voltaian comprises the Oti and the Obosum beds, which are well consolidated and generally flat lying. The beds are made up of interbedded mudstones/siltstones, sandstones, arkoses, and conglomerates [15]. Details of geology and hydrogeology are documented in detailed reports and articles on the basin [14, 16–21]. The rocks are largely impervious, so that the occurrence of groundwater is associated with the development of secondary porosity through jointing, shearing and fracturing, and weathering. Rock compaction and slight metamorphism is believed to have destroyed the primary porosity [16], leading to considerably reduced inherent primary permeabilities. However, where intense weathering occurs, the rocks serve as better aquifers with significantly enhanced hydraulic and storage properties. The nature, aperture, and degree of interconnection between joints determine the hydrogeological fortunes of the rocks [14, 17–19]. Drilling projects and hydrological investigations reveal that shallow potential aquifers capable of delivering water of sustainable quantities for domestic and industrial use exist in the basin [20].

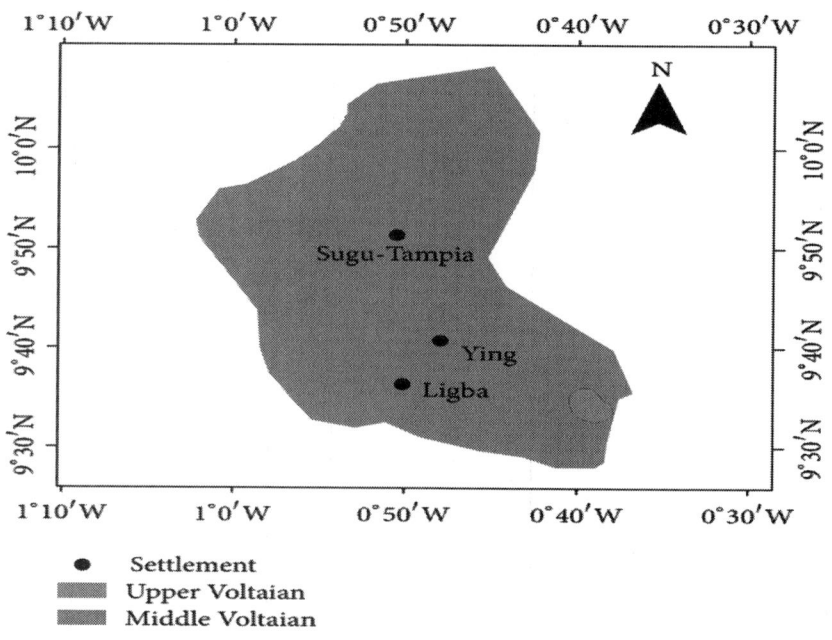

Figure 2: A geological map of the study area.

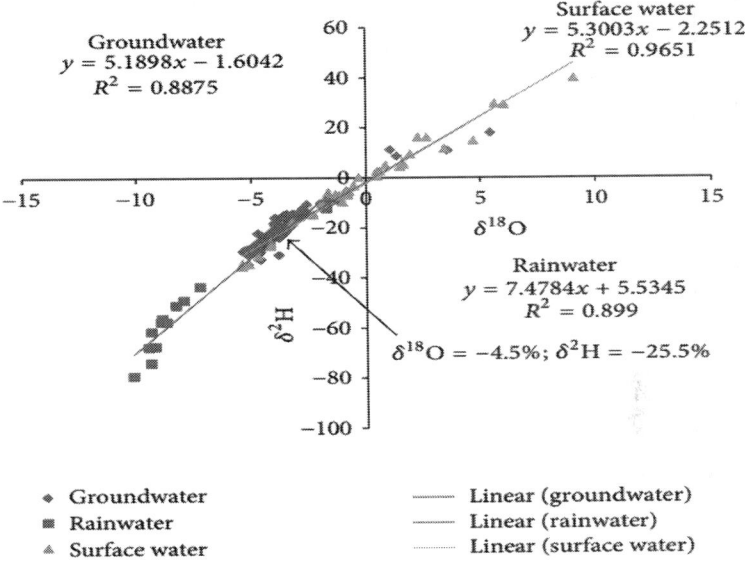

Figure 3: Stable isotope signatures of rainwater, surface water, and groundwater from parts of the White Volta Basin, Ghana.

Peasant rain-fed agriculture is the main source of employment in the basin. However, recent erratic patterns of rainfall in the area, coupled with rising populations in the basin, have led to increased interests in developing the irrigation potentials in the area. This has heightened interests in assessing the potentials of developing the aquifers in the area for commercial abstraction to support irrigation activities to enhance food security and reduce poverty since agriculture is the mainstay of the communities there.

MATERIALS AND METHODS

A total of forty-two samples (19 groundwater, 11 rainwater, and 12 surface water samples) were collected for the study. Rainwater was taken in Tamale and analyzed for both the chloride and isotopes (^{18}O and ^2H). Historical groundwater hydrochemical data from the basin were obtained from the Community Water and Sanitation Agency (CWSA) in the Northern Region. The surface water samples were taken from tributaries of the White Volta. Strict adherence to sampling protocols was followed for collected water samples both in the field and the laboratory. All the analyses were carried out at the Ghana Atomic Energy Commission's chemistry laboratory. Chloride analysis was done using a Dionex ICS 90 ion chromatograph equipped with an AS14A-5 µm ion pac column. Stable isotope analysis of $\delta^{18}O$ was carried out using VG Sira 10 mass spectrometry. Delta deuterium measurements were done on a EuroVector elemental analyzer (EA; EuroPyrOH-3100) with a liquid autosampler (LAS; Euro AS-300) coupled to a Micromass IsoPrime isotope ratio mass spectrometer. The isotope composition of water is reported as the deviation of $^2H/^1H$ or $^{18}O/^{16}O$ ratio from that of Vienna Standard Mean Ocean Water (VSMOW) in parts per thousand (%$_0$).

The CMB methodology is summarized in

$$\text{Recharge} = \frac{C_p}{C_{gw}} P,$$

(1)

where C_p and C_{gw}, respectively, represent chloride concentrations in precipitation and groundwater and P represents the average annual depth of precipitation in the area.

Average chloride concentration in rainwater was obtained from analyses of rainwater samples taken in Tamale, the regional capital of the Northern Region, as part of this study. Groundwater chloride concentration was obtained from the historical hydrochemical data that had already been procured from the CWSA. An average annual precipitation rate of 1100 mm was used in the estimation. Ordinary Kriging was then applied to the estimated recharge to achieve a regional distribution in the area. Ordinary kriging is a linear estimation method which is based on [22].

$$Z(x) = \sum_{i=1}^{n} \lambda_i Z(x_i),$$

(2)

where λ_i refers to the weight assigned to a known or estimated value $Z(x_i)$, of the parameter, and $Z(x)$ is the new value being estimated.

Different estimation methods use different criteria to assign the weights to parameters, but most of them are based on the proximities of the known locations to the locations being estimated [22]. Where proximity is the sole determinant, the weights are assigned such that closer points have higher contributions to the estimates than farther points as is the case with inverse distance weighting [22]. The advantage of ordinary kriging over the others is that it uses both the proximity factor and the general pattern of spatial variation of the parameter with distance. A variogram is modeled from the original data as part of the processes of estimating the weighting factors or coefficients. Such a model characterizes the spatial dependence of the parameter in the domain. It is the fitted experimental model which is then used together with the proximity factor to estimate coefficients for all the data to be used in the estimation.

In this study, both the variography and ordinary kriging were performed from the R-platform. An acceptable variogram was carefully chosen and fitted to the computed dataset of the recharge estimates. This was achieved by trying the various possible theoretical models available until the most appropriate model was achieved. Ordinary kriging was then performed from the variogram model chosen. Prediction accuracy using ordinary kriging and any kriging methodology partly hinges on how adequately the spatial dependence of the parameter was modeled. This implies that the choice of an inappropriate model of spatial dependence can potentially lead to inaccurate estimates or predictions.

RESULTS AND DISCUSSIONS

The sources and origin of groundwater recharge in the Voltaian was assessed using stable isotope data of precipitation (rainfall), groundwater, and surface water from parts of the Voltaian. Historical stable isotope data ($\delta^{18}O$ and $\delta^{2}H$) for groundwater, surface water, and rainfall were plotted on a biplot for the purposes of ascertaining the sources and/or evolution of groundwater in the area. Isotope tracers are amongst the most frequently used to trace the sources and/or genesis of water reservoirs and contaminants and have been noted to provide useful insights to guide further detailed investigations. The ratio of the rarer (and most often the heavier) isotope to the more abundant (often the lightest) isotope provides an indication of the relative enrichment of the two isotopes in the medium or the original source of recharge. It is such a ratio that provides indications of the climatic conditions prevailing at the period and location of recharge and can therefore be used qualitatively to infer the age of a water body. These ratios are expressed in relation to an international standard which provides some uniformity for comparing environments or reservoirs on a global scale and is often expressed in the delta (δ) notation as indicated in

$$\delta^{18}O = \left(\frac{\left({}^{18}O / {}^{16}O\right)_{sample}}{\left({}^{18}O / {}^{16}O\right)_{V\text{-}SMOW}} \right) \times 1000,$$

$$\delta^{2}H = \left(\frac{\left({}^{2}H / {}^{1}H\right)_{sample}}{\left({}^{2}H / {}^{1}H\right)_{V\text{-}SMOW}} \right) \times 1000,$$

$$(3)$$

where the terms in the numerator and denominator, respectively, represent the ratio of the heavier to the lighter isotope in the sample and international standard, respectively.

The global meteoric water line (GMWL) was first published by Craig [23] and is a convenient reference for understanding and tracing water origin. It is a linear relation in the form of

$$\delta^{2}H = 8\delta^{18}O + d,$$

$$(4)$$

where d, the y-intercept, is the deuterium excess (or d-excess) parameter when the slope = 8 [24]. From Craig's MWL, d=10 at this slope and is indicative of no evaporative effect during precipitation. The underlying assumption [23] is that water with an isotopic composition that falls along the GMWL originates from the atmosphere and is relatively unaffected by isotopic processes. Isotopic signatures from different reservoirs are often discussed in relation to (4). In this study, groundwater recharge in the area appears to be of meteoric origin as the groundwater data plots close to the GMWL, albeit with shallower slope and intercept, which suggest some degree of enrichment of the heavier isotope relative to the lighter ones (Figure 3). The observed pattern is consistent with conditions of lower relative humidity than 100% and high ambient temperatures as is common in the study area. The shallower slope and d-excess values result from evaporation of raindrops due to low relative humidities during the course of the rains. The average relative humidity in the area during the rainy season is about 70% [13]. Vertical percolation of precipitation down to the saturated zone is variable in the space of the study area, due to the variability in the clay content which limits the vertical hydraulic conductivity of the material of the unsaturated zone. Where the clay content is high, vertical percolation is significantly restricted, and the resulting recharged groundwater is significantly enriched due to evaporative effects of high temperatures and low humidities. It is obvious that the surface water data fall exactly on the Global Meteoric Water Line, GMWL, whereas both the rainwater and groundwater samples are relatively enriched, with shallower slopes and deuterium excess (d-excess) values (Figure 3). The rainwater samples were taken during three events in the major rainy season and may not adequately reflect the signature for the entire year. However, studies conducted in the southern parts of the basin using data mainly of the dry season precipitation suggest significant enrichment [25]. A solution of the equation for the Local Meteoric Water Line (LMWL) developed from the precipitation data and the local Groundwater Line (LGWL) developed from the groundwater isotope data provides the isotopic signature of the most probable source precipitation of the groundwater in the area. The intersection of the LGWL and LMWL in Figure 3 provides such a signature. As is obviously presented in Figure 3, the isotopic signature at the point of intersection of these two lines ($\delta^{18}O = -4.5$; $\delta^2H = -25.5$) suggests that the source of recharge is recent meteoric water or a mixture of precipitation types which are of recent origin. The local surface water line (LSWL) is similar to the groundwater line in terms of slope and intercept (Deuterium excess) and suggests that both have a similar source and may have been affected by similar processes over time. A recent assessment of the isotope characteristics of the entire Voltaian indicated a similar isotopic signature of the source water of groundwater recharge in the basin [26] and indicated that the processes of infiltration and percolation of precipitation water may be variably slow in the terrain, due to the variability in the nature of the overburden even within short distances. Yidana [26] estimated the rates of evaporation of precipitation water in transit down the unsaturated zone to the saturated zone and reports evaporative losses of 19.8%–70.6% for the Voltaian sedimentary basin in Ghana. This contrasts with significantly higher evaporation rates estimated for surface water bodies in the area (29.5%–84.7%) and highlights the fact that surface water, in view of its continuous exposure to the atmosphere endures higher impacts of higher ambient temperatures and low humidities than infiltrating water. In both cases, however, the estimates are most likely the average conditions over a long period of time since the waters sampled are most likely mixed waters from different events. These estimates do not include the effects of transpiration as the impacts of transpiration on isotopic signature of water are difficult to estimate. However, transpiration contributes significantly to water loss especially in the unsaturated zone. Therefore, if included, the rate of water loss due to the combined effects of evaporation and precipitation in the entire terrain and in the study area will be significantly higher.

Estimated groundwater recharge from the CMB suggests annual recharge in the range of 0.9%–21% of the total annual precipitation. This is consistent with the observation of high evaporation rates estimated for the entire basin and suggests that much more water may have been lost to transpiration. However, this much of groundwater recharge suggests high fortunes in terms of commercial groundwater resources development in the area, if a significant proportion of it is available for abstraction. The wide range in the estimated data and the high standard deviation suggests significant spatial variation in groundwater recharge rates in the study area. This may be related to the nature of the unsaturated zone material and its variability in the space of the domain of this study. The average rate of about 5.5% is significant and compares favorably estimates in other parts of the terrain. For instance Attandoh et al. [11] estimated groundwater recharge in the range of 0.3%–4.1% of total annual precipitation in the area, through model calibration in the same terrain. Earlier estimates of Yidana [27] in the south of the Voltaian suggested similar rates through model calibration.

Ordinary kriging was employed to predict the spatial pattern of variation in groundwater recharge in the study area. Figure 4 presents the variogram that was fitted to the original, estimated groundwater recharge rates from the CMB methodology. A spherical variogram was adjudged appropriate for the distribution and was accordingly chosen for the estimation stage of ordinary kriging. This ensured that the prediction is as accurate as possible. The resulting predicted surface for the spatial variation in recharge in the domain is presented in Figure 5. A significant spatial pattern of variability in groundwater recharge is suggested in Figure 5.

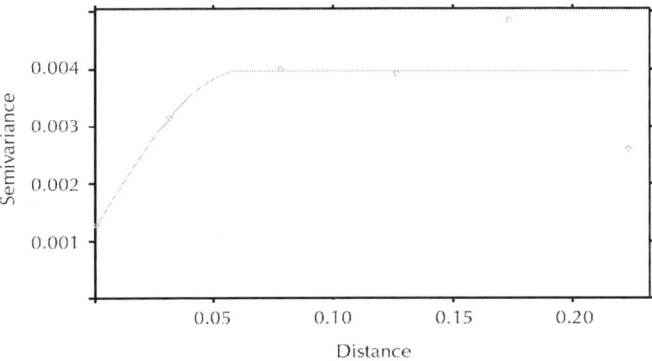

Figure 4: Semivariogram for estimated groundwater recharge in the study area.

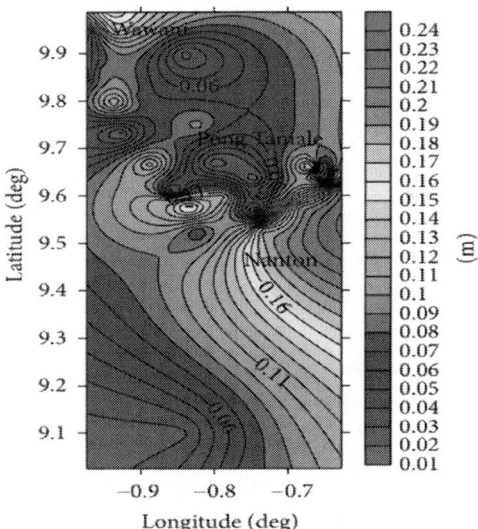

Figure 5: Spatial distribution of estimated groundwater recharge in Savelugu and surrounding subcatchments of the White Volta Basin.

There is an apparently high variability in direct groundwater recharge from precipitation in the area and much of the Voltaian basin. This observation is consistent with the nature of the material in the unsaturated zone which varies in space in terms of the clay content.

Where the clay content is considerably high, vertical percolation of rainwater is much reduced, leading to reduced vertical recharge. Infiltrating rainwater experiencing such restricted vertical flow therefore undergoes significant evaporation such that a high percentage is lost to the atmosphere. Yidana [26] suggests that, on the basis of the analyses of stable isotope data of precipitation, rainwater, and groundwater from parts of the entire basin, evaporation of infiltrating rainwater is in the range of 19.8% and 70.6% of the total annual precipitation in the basin. This is quite consistent with the observed low humidities and high annual temperatures and explains why surface impoundments have not been successful as sustainable sources of irrigation water in the region.

The estimated groundwater recharge distribution appears to be consistent with configuration of the groundwater flow pattern suggested by the groundwater table map (Figure 6) and the surface topographical map (Figure 7) of the area. The highest recharge areas (Figure 5) are generally in the eastern parts of the area, which coincide with the local recharge areas suggested by the potentiometric map (Figure 5).

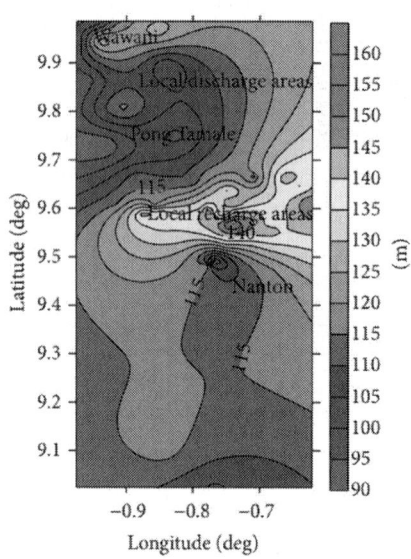

Figure 6: Potentiometric map of the study area showing local recharge and discharge areas.

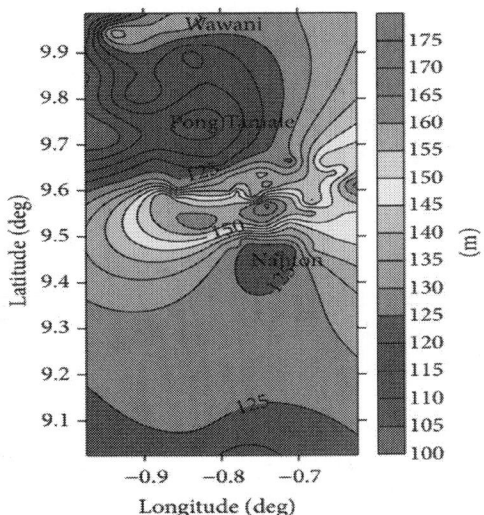

Figure 7: Local topographical map of the study area.

Some of the areas of low recharge in the north and south of the study area (Figure 5) also coincide with locations of the lowest hydraulic head (Figure 6) and suggests that the groundwater recharge rates from the CMB methodology accurately accounts for the observed groundwater flow geometry in the area. Groundwater recharge estimates from the CMB methodology is largely accurate when applied in arid to semiarid areas where local surface runoff is negligible [28]. Yidana and Koffie [29] applied the same methodology to similar aquifers in the north of the area and suggested groundwater recharge in the range of 1.8% and 32% of the total annual rainfall in the area. These variable recharge rates predicted in their study are consistent with the nature of the material in the unsaturated zone as suggested by some well logs in the area.

Estimating one of the key hydrogeological parameters is based on a proper understanding of estimates of other uncertain parameters in the terrain in hydrogeological modeling. In the study of Attandoh et al. [11], the key parameters of aquifer hydraulic conductivity and recharge were estimated through model calibration. Previous researchers [24, 30] criticized the estimation of key parameters of groundwater flow in this pattern, contending that a wide range of values could produce the same level of calibration, leading to the lack of uniqueness in such estimates. The range of estimates obtained from the current study provides a wider range and probably accounts for the marked variability in the nature of the material in the unsaturated zone, which regulates groundwater recharge. In addition, as suggested in the study of Attandoh et al. [11] much of the recharge was computed through the general head boundary. In effect, when all the components of recharge estimated from the model are put together, the range is compatible with those computed with the CMB methodology in this study. This suggests that, in spite of the criticisms of nonuniqueness in parameter estimates through model calibration, the methodology provides quite reliable estimates if carefully executed.

CONCLUSIONS

Results of this study suggest that the chloride mass balance (CMB) approach performs well in a tropical setting in providing fairly accurate estimates of groundwater recharge for groundwater resources evaluation. Estimates in this study compare well with estimates from numerical model calibration and analyses of evaporation rates of infiltrating rainwater. Estimated groundwater recharge from the CMB indicates that 0.9% to over 21% of annual precipitation recharges the shallow aquifer system in the area. This much of groundwater recharge suggests high fortunes in terms of commercial groundwater resources development in the area. The study also indicates that the source of groundwater recharge in the terrain is recent meteoric water which may have been isotopically enriched in the heavier isotopes of the elements of the water molecule during the process of recharge. It is suggested in this study that, whenever the CMB methodology must be used in groundwater recharge estimation, other tracers must be used to check the validity of the results as indicated in this study.

REFERENCES

1.   C.-S. Ting, T. Kerh, and C.-J. Liao, "Estimation of groundwater recharge using the chloride mass-balance method, Pingtung Plain, Taiwan," Hydrogeology Journal, vol. 6, no. 2, pp. 282–292, 1998.

2. D. N, Lerner, A. S. Issar, and I. Simmers, Groundwater Recharge, A Guide to understanding and Estimating Natural Recharge, UNESCO International Hydrogeological Program, International Association of Hydrogeologists, Hannover, Germany.

3. W. C. Fetter, Applied Hydrogeology, Prentice Hall, Upper Saddle River, NJ, USA, 4th edition, 2000.

4. R. W. Healy and P. G. Cook, "Using groundwater levels to estimate recharge," Hydrogeology Journal, vol. 10, no. 1, pp. 91–109, 2002.

5. A. Horst, J. Mahlknecht, B. J. Merkel, R. Aravena, and Y. R. Ramos-Arroyo, "Evaluation of the recharge processes and impacts of irrigation on groundwater using CFCs and radiogenic isotopes in the Silao-Romita basin, Mexico," Hydrogeology Journal, vol. 16, no. 8, pp. 1601–1614, 2008.

6. L. Dassi, "Use of chloride mass balance and tritium data for estimation of groundwater recharge and renewal rate in an unconfined aquifer from North Africa: a case study from Tunisia," Environmental Earth Sciences, vol. 60, no. 4, pp. 861–871, 2010.

7. A. M. Subyani, "Use of chloride-mass balance and environmental isotopes for evaluation of groundwater recharge in the alluvial aquifer, Wadi Tharad, western Saudi Arabia," Environmental Geology, vol. 46, no. 6-7, pp. 741–749, 2004.

8. A. S. Bazuhair and W. W. Wood, "Chloride mass-balance method for estimating ground water recharge in arid areas: examples from western Saudi Arabia," Journal of Hydrology, vol. 186, no. 1-4, pp. 153–159, 1996.

9. S. W. West and W. L. Broadhurst, "Summary appraisals of Nation's ground-water resources—Rio Grande region," US Geological Survey Professional Paper 813 D.

10. W. W. Wood and W. E. Sanford, "Chemical and isotopic method for quantifying ground—water recharge in a regional, semiarid environment," Ground Water, vol. 33, no. 3, pp. 456–486, 1995.

11. N. Attandoh, S. M. Yidana, A. Abdul-Samed, P. A. Sakyi, B. Banoeng-Yakubo, and P. Nude, "Conceptualization of the hydrogeological system of some sedimentary aquifers in Savelugu-Nanton and surrounding areas, Northern Ghana," Hydrological Processes, vol. 27, pp. 1664–1676, 2013.

12. F. A. Armah, D. O. Yawson, G. T. Yengoh, J. Odoi, and E. K. A. Afrifa, "Impact of floods on livehoods and vulnerability of natural resource dependent communities in Northern Ghana," Water, vol. 2, no. 2, pp. 120–139, 2010.

13. K. Dickson and G. Benneh, A New Geography of Ghana, Longmans Group Limited, London, UK, 5th edition, 2004.

14. S. Y. Acheampong and J. W. Hess, "Origin of the shallow groundwater system in the southern Voltaian Sedimentary Basin of Ghana: an isotopic approach," Journal of Hydrology, vol. 233, no. 1-4, pp. 37–53, 2000.

15. G. O. Kesse, The Mineral and Rock Resources of Ghana, A. A. Balkema, Rotterdam, The Netherlands, 1985.

16. S. Dapaah-Siakwan and P. Gyau-Boakye, "Hydrogeologic framework and borehole yields in Ghana,"Hydrogeology Journal, vol. 8, no. 4, pp. 405–416, 2000.

17. N. R. Junner and H. Service, "Geological notes on Volta River District and Togoland under British mandate," Annual Report on the Geological Survey by the Director, 1935.

18. N. R. Junner and T. Hirst, The Geology and Hydrogeology of the Volta Basin, vol. 8 of Memoir, Gold Coast Geological Survey, The Gold Coast, Australia, 1946.

19. S. M. Yidana, D. Ophori, and B. Banoeng-Yakubo, "Hydrogeological and hydrochemical characterization of the Voltaian Basin: the Afram Plains area, Ghana," Environmental Geology, vol. 53, no. 6, pp. 1213–1223, 2008.

20. B. Banoeng-Yakubo, S. Yidana, J. Ajayi, Y. Loh, and D. Asiedu, "Hydrogeology and groundwater resources of ghana: a review of the hydrogeological zonation in Ghana," in Potable Water and Sanitation, J. M. McMann, Ed., Nova Science, 2010.

21. S. Y. Achcampong and J. W. Hess, "Hydrogeologic and hydrochemical framework of the shallow groundwater system in the southern Voltaian Sedimentary Basin, Ghana," Hydrogeology Journal, vol. 6, no. 4, pp. 527–537, 1998.

22. E. H. Isaaks and R. M. Srivastava, Applied Geostatistics, Oxford University Press, Oxford, UK, 1989.

23. H. Craig, "Isotopic variations in meteoric waters," Science, vol. 133, no. 3465, pp. 1702–1703, 1961.

24. L. F. Konikow and J. D. Bredehoeft, "Ground-water models cannot be validated," Advances in Water Resources, vol. 15, no. 1, pp. 75–83, 1992.

25. S. Y. Acheampong and J. W. Hess, "Origin of the shallow groundwater system in the southern Voltaian Sedimentary Basin of Ghana: an isotopic approach," Journal of Hydrology, vol. 233, no. 1-4, pp. 37–53, 2000.

26. S. M. Yidana, "Stable isotope characteristics of groundwater in the voltaian basin: an evaluation of the role of meteoric recharge in the basin," Journal of Hydrogeology & Hydrologic Engineering, vol. 2, no. 2, pp. 1–10, 2013.

27. S. M. Yidana, "Groundwater flow modeling and particle tracking for chemical transport in the southern Voltaian aquifers," Environment Earth Science, vol. 63, pp. 709–721, 2013.

28. N. L. Nyagwambo, Groundwater recharge estimation and water resources assessment in a tropical crystalline basement aquifer [PhD dissertation], UNESCO-IHE Institute for Water Education, Delft, The Netherlands, 2006.

29. S. Yidana and E. Koffie, "The groundwater recharge regime of some slightly metamorphosed neoproterozoic sedimentary rocks: an application of natural environmental tracers," Hydrological Processes, vol. 28, no. 7, pp. 3104–3117, 2013.

30. C. R. Jackson, "Hillslope infiltration and lateral downslope unsaturated flow," Water Resources Research, vol. 28, no. 9, pp. 2533–2539, 1992.

Forecasting the Development of Boreal Paludified Forests in Response to Climate Change: A Case Study Using Ontario Ecosite Classification

Benoit Lafleur[1, 2], Nicole J Fenton[1], and Yves Bergeron[1]

[1]Institut de recherche sur les forêts, Université du Québec en Abitibi-Témiscamingue, 445 boul. de l'Université, Rouyn-Noranda J9X 5E4, QC, Canada

[2]Centre d'étude de la forêt, Université du Québec à Montréal, 141 Avenue du Président-Kennedy, Montréal H2X 1Y4, QC, Canada

ABSTRACT

Background

Successional paludification, a dynamic process that leads to the formation of peatlands, is influenced by climatic factors and site features such as surficial deposits and soil texture. In boreal regions, projected climate change and corresponding modifications in natural fire regimes are expected to influence the paludification process and forest development. The objective of this study was to forecast the development of boreal paludified forests in northeastern North America in relation to climate change and modifications in the natural fire regime for the period 2011–2100.

Methods

A paludification index was built using static (e.g. surficial deposits and soil texture) and dynamic (e.g. moisture regime and soil organic layer thickness) stand scale factors available from forest maps. The index considered the effects of three temperature increase scenarios (i.e. +1°C, +3°C and +6°C) and progressively decreasing fire cycle (from 300 years for 2011–2041, to 200 years for 2071–2100) on peat accumulation rate and soil organic layer (SOL) thickness at the stand level, and paludification at the landscape level.

Results

Our index show that in the context where in the absence of fire the landscape continues to paludify, the negative effect of climate change on peat accumulation resulted in little modification to SOL thickness at the stand level, and no change in the paludification level of the study area between 2011 and 2100. However, including decreasing fire cycle to the index resulted in declines in paludified area. Overall, the index predicts a slight to moderate decrease in the area covered by paludified forests in 2100, with slower rates of paludification.

Conclusions

Slower paludification rates imply greater forest productivity and a greater potential for forest harvest, but also a gradual loss of open paludified stands, which could impact the carbon balance in paludified landscapes. Nonetheless, as the thick *Sphagnum* layer typical of paludified forests may protect soil organic layer from drought and deep burns, a significant proportion of the territory has high potential to remain a carbon sink.

BACKGROUND

In boreal forest ecosystems, successional paludification is described as a dynamic process driven by forest succession between fire events that leads to peat accumulation, and a concomitant thickening of the soil organic layer (SOL), and the formation of waterlogged conditions on a formerly dry mineral soil (Simard et al. [2007]). Paludification is influenced by climatic factors and permanent site features, such as surficial deposits and soil texture, as well as by natural fire regimes (Lecomte et al.[2006]; Simard et al. [2009]; Payette et al. [2013]). In boreal regions, in the extended absence of fire, paludification leads to the formation of paludified forests and can reduce forest productivity by up to 50%–80% (Simard et al. [2007]).

According to the most recent report of the Intergovernmental Panel on Climate Change, warming of the climate system is unequivocal (IPCC [2013]). This changing climate is expected to increase drought severity in boreal regions (Girardin and Mudelsee [2008]), and therefore to influence the natural fire regime, resulting in an increase of fire severity and burn rate (Flannigan et al. [2005]; de Groot et al. [2009]; Bergeron et al. [2010]; van Bellen et al. [2010]). Fire plays an important role in landscape level paludification processes, as fire can "depaludify" forest stands if most of the SOL is burnt (Dyrness and Norum [1983]; Greene et al. [2005]). However, if the fire is not severe and a relatively thick SOL remains after fire, the regenerating forest stands may remain paludified (Lecomte et al. [2005]). Because boreal peatlands represent important carbon reservoirs (it is estimated that boreal peatlands, including paludified forests, store 455 Pg of carbon, i.e. approximately 15% of the Earth's terrestrial carbon (Gorham

[1991]; Lavoie et al. [2005])) any modification to the fire cycle may have important consequences on the carbon cycle and the global climate.

Throughout the boreal region, paludified forests support important forest industries. It is expected that any modifications to the climate and natural fire regimes will, in all likelihood, require industries to adapt to new ecosystem conditions and, presumably, to modify their practices. Depending on the type of harvest practices, forest harvest in paludified forests can both promote or reduce paludification (Lafleur et al. [2010a], [2010b]), and, as is the case for fire, have important effects on the C budget at the landscape level. This context provides strong incentives for the development of simple tools that can be used to rapidly and easily forecast the combined effects of climate change and modified fire regime on paludification and forest development.

In North America, forest mapping is commonly used to describe the forest mosaic at the regional scale. Forest maps provide information on stand scale environmental factors (e.g. surficial deposits, soil texture, moisture regime, slope), as well as on stand species composition, height and density. This information presents a great potential for research in forest ecology and management. For instance, it can be used by forest managers to forecast the effects of silvicultural practices or wildfire on stand regeneration, composition and productivity. Because some stand scale environmental factors provided in forest maps are intrinsically dynamic (e.g. soil moisture regime and SOL thickness) and potentially influenced by climate variables, forest maps could also be used to forecast the effects of climate change on forest development.

In this context, the main objective of this study was to use information commonly available on forest maps in Canada in order to evaluate the potential for paludification of boreal forest stands in relation to climate change and modifications to the natural fire regime. To achieve this objective we developed a dynamic paludification index. First, we developed a base index projecting the development of paludified forests over time without considering climate change. Then, we added the effects of climate change on the thickening on SOL to this base index in order to forecast the development of paludified forests in response to projected climate change. Finally, forecasted climate change and modifications to the natural fire regime were added to the base index to further explore the development of paludified forests in the context of climate change. Based on information commonly available on forest maps, the paludification index can therefore estimate the effects of both climate change and natural fire regimes on the development of paludified forests at both the local and regional scales.

METHODS

Study Area

A territory in eastern Canada was used to model the development of paludified forests in relation to climate change and changes in the natural fire regime. The Gordon Cosens Forest is a 20 000 km²(17 360 km² of which is covered in forest, and the remaining area is water bodies) forest management unit located in the Ontario ecodistrict 3E-1, a region also known as the Clay Belt (Figure 1). Cold climate, flat topography, and surficial deposits that are resistant to water penetration all make this region favorable for the development of paludified forests (Jeglum [1991]; Riley[1994]). The southern part of the Clay Belt is covered by thick (>10 m) glaciolacustrine clay and silt deposited by glacial Lake Ojibway, while the northern part, known as the Cochrane till, is covered by a compact till made up of a mixture of clay and gravel, created by a southward ice flow approximately 8000 years BP (Veillette [1994]). Soils of the study area are mostly classified as Gleysols and Luvisols (Soil Classification Working Group [1998]). Nonetheless, organic deposits (i.e. a surficial deposit consisting of a SOL > 40 cm thick) are found in many locations in both the southern and northern parts of the study area. Black spruce (*Picea mariana* [Mill.] BSP.) is the dominant tree species of the study area.

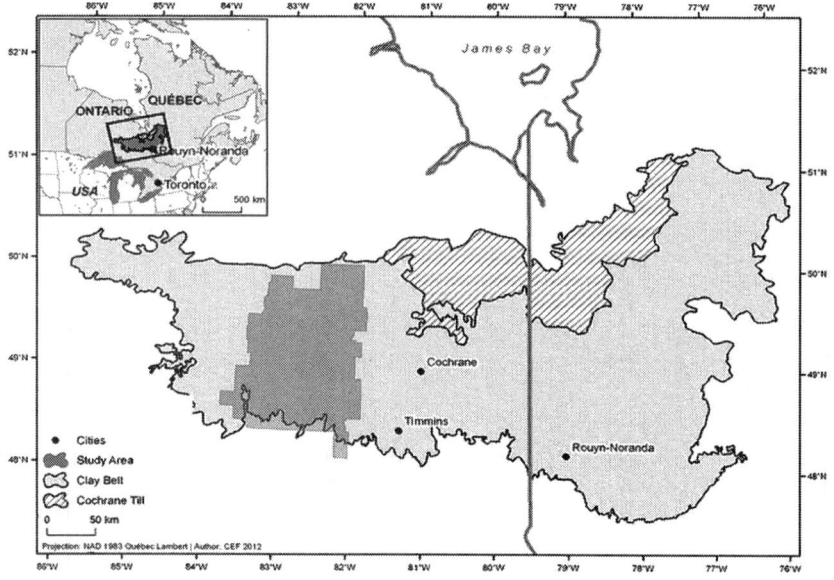

Figure 1: Location of the Clay Belt and the Gordon Cosens Forest in northeastern Ontario (inset).

According to the local weather station (Kapuskasing, Ontario), from 1981 to 2010, the average annual temperature was 1.3°C and the average annual precipitation was 830 mm, with 30% falling during the growing season (Environment Canada [2014]). The average number of degree-days (>5°C) is 1430, and the frost-free season lasts about 100 days; frost can occasionally occur during the growing season. In this region, according to different scenarios (A1 and B2) and simulation models (Canadian Center for Climate Modelling and Analysis [CGCM 3.1], Australian-based Commonwealth Scientific and Industrial Research Organization [CSIROMk3.5], National Institute for Environmental Studies [MIROC 3.2 medres], and National Center for Atmospheric Research [NCAR-CC SM3]), climate warming is projected to be between 3–6°C and precipitation is projected to increase by 10%–20% by the end of the 21st century (McKenney et al. [2010]).

In the study area, fire frequency has diminished from a 100-year cycle to an approximately 400-year cycle since the little Ice Age (ca. 1850; Bergeron et al. [2004]). As a result of anthropogenic climate change, Bergeron et al. ([2010]) predict a doubling of fire frequency in this region by the end of the 21st century.

Map Data

Data for individual forest stands polygons, used to forecast the development of boreal forested peatlands, were retrieved from the Ontario ecosite classification system (Taylor et al. [2000]). This classification seeks to classify the province's ecosystems, such as non forested uplands (prairie, cliff top, dunes), forested ecosystems (both upland and lowland), and non forested wetlands (marshes, swamps, fen, bogs). For forest polygons, ecosites are defined as homogeneous landscape areas (i.e. polygons typically 10–100 hectares) of common surficial deposits, soil moisture regime, soil texture, SOL thickness, humus form, and tree cover. Figure 2 illustrates the forest stand mosaic of the Gordon Cosens Forest, whereas Table 1 shows the size distribution of forest stand polygons. Ecosite classification is meant to be a practical tool for resource managers, and can be used for a variety of forest and site level applications including timber supply analysis, harvest planning, wildlife habitat studies and assessments, and successional studies.

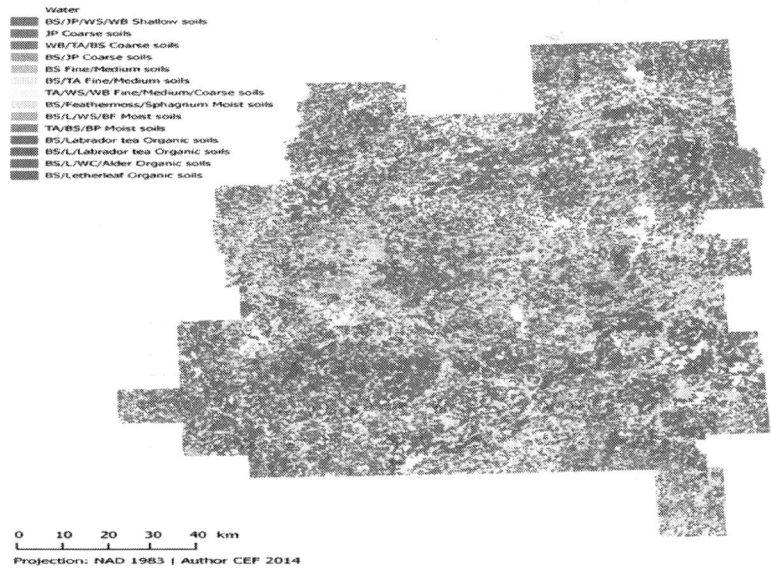

Figure 2: Forest mosaic of the Gordon Cosens Forest. BF = balsam fir, BP = balsam poplar, BS = black spruce, JP = jack pine, L = larch, TA = trembling aspen, WB = white birch, WC = white cedar, WS = white spruce.

Table 1: Size and area distribution of polygons of the Gordon Cosens Forest map

Polygon area (ha)	Number of polygons	Total area (km²)
<1	9402 (11%)	29 (0.2%)
1–10	36030 (41%)	1702 (9.8%)
10–100	38694 (45%)	11631 (67.0%)
>100	2489 (3%)	1702 (23.0%)

Lafleur et al.

Lafleur et al. Forest Ecosystems 2015 2:3, doi:10.1186/s40663-015-0027-6

The forest stand polygon data used in this study were retrieved from the 2004 forest inventory reassessment, the most recent reassessment available for the study area. The analysis of this inventory reveals that in 2004, organic deposits (i.e. surficial deposit consisting of a SOL > 40 cm thick) covered 42% of the study area, and nearly 70% of the area had moist to wet soils. While black spruce stands covered 55% of the study area, trembling aspen (*Populus tremuloides* Michx.) and mixed black spruce-trembling aspen stands covered 5% and 25% of the area, respectively. Jack pine (*Pinus banksiana* Lamb.), white spruce (*Picea glauca* Moench), and white birch (*Betula papyrifera*Marsh.) were the other dominant species of the regional landscape.

Development of the Paludification Index

Using static (e.g. surficial deposit and soil texture) and dynamic (e.g. moisture regime and SOL thickness) stand scale factors available from forest maps (Table 2; Taylor et al. [2000]), we developed a dynamic paludification index for each forest stand polygon. Each static and dynamic factor was divided into classes (2 to 9 according to site feature), each of which was attributed a score related to its paludification "power" (between 0 and 5; 0 = null paludification "power" and 5 = high paludification "power"). Table 2 lists the factor classes with their related paludification "power" score. Adding up the score of each factor, gives the paludification index, which describes the level of paludification of each forest stand polygon (hereafter referred to as stand). The maximum score an ecosite could achieve is 14. The paludification index was calculated as follows:

$$PI = \text{Static factors} + \text{Dynamic factors}$$

(1)

or

$$PI = (D + T + H) + (SOL + M + O),$$

(2)

where PI is the paludification index, D is surficial deposit, T is soil texture, H is humus form, SOL is soil organic layer thickness, M is moisture regime, and O is overstorey composition. Although we recognize the dynamic nature of surficial deposits and soil texture, we treated these factors as static because the transformation rates of these variables are slow relative to the projection time (i.e. 100 years) used in this study. Similarly, humus form was considered to be static because of its slow rate of transformation (Yu et al. [2001]). Stands with a $PI \geq 13$ were classified as *paludified*, whereas stands with a PI between ≥ 7 and < 13 were classified as *nearly paludified*. Stands with a $PI \leq 6$ were classified as *not paludified*. Although we recognize the possibility of interactions among variables and nonlinearity in the effects, the use of a simple index with linear effects was justified because of the lack of information about possible interactions among some variables.

Table 2: Paludification scores for static and dynamic stand scale factors

Surficial deposit (D)		Soil texture (T)		Humus form (H)		Soil organic layer thickness (SOL)		Moisture regime (M)		Overstorey (O)	
Class	Score	Class[1]	Score	Class[1]	Score	Class[2]	Score	Class[3]	Score	Class	Score
Rock	0	Rock	0	Mull	0	0–9	0	Dry	0	Other spp.	0
Aeolian	1	Sandy	0	Moder	0	10–19	1	Moderately fresh	0	Black spruce	1
Fluviatil	1	Coarse loam	0	Humic mor	1	20–29	1	Fresh	0		
Fluvial till	1	Medium loam	1	Fibric mor	1	30–39	2	Very fresh	0		
Clay till	2	Silty	1	Humic	2	40–120	2	Moderately moist	1		
Lacustrine	2	Fine loam	1	Mesic	2	>120	2	Moist	2		
Organic	2	Clay	2	Fibric	2			Very moist	3		
								Wet	4		
								Very wet	5		

Each static and dynamic factor was divided into classes, each of which was attributed a score related to its paludification "power"; 0 = null "power", 1 = low "power", 5 = high "power". Adding up the score of each factor, gives a Paludification Index that estimates the liability of each stand to paludification.

[1]According to Sims and Baldwin ([1996]).

[2]In cm. In our paludification index, we used the median value of each class.

[3]According to Sims et al. ([1989]).

Lafleur et al.

Lafleur et al. Forest Ecosystems 2015 2:3, doi:10.1186/s40663-015-0027-6

Dynamic factors were allowed to vary over time according to certain rules. In the base index (i.e. the index not considering climate change or fire), SOL thickness increased with time, following peat accumulation rates determined by Lecomte et al. ([2006]) for nearby sites located in the Clay Belt of Quebec. According to Lecomte et al. ([2006]), for the past ca. 200 years the peat accumulation rate varied between 10 and 20 cm per century. Lecomte et al. ([2006]) also showed that peat tends to accumulate at a faster rate where SOL is > 20 cm deep. As a result, the base index was adjusted to allow stands with a median SOL depth > 20 cm to accumulate peat at a rate of 20 cm per century, whereas stands with SOL median depth < 20 cm we allowed to accumulate peat at a rate of 10 cm per century. In the same vein, stands with an initial SOL depth < 20 cm had their peat accumulation rate adjusted to 20 cm per century when their SOL depth reached 20 cm. Simard et al. ([2009]) and Drobyshev et al. ([2010]) observed an important decline in tree growth when SOL depth is 20 cm or greater, further confirming the pertinence of the 20 cm cut off point. Furthermore, in the base index, moisture regime was allowed to vary in stands where soil texture was finer than medium loam, only if SOL depth was ≥ 20 cm. Hence, in all cases when SOL depth reached ≥ 20 cm, moisture regime stepped one class ahead. For example, stands with a fine loam soil texture and a moisture regime classified as *Moist* had their moisture regime changed to *Very moist* when their SOL depth reached 20 cm.

In the index considering climate change, peat accumulation rate (hence SOL thickness) was allowed to vary according to an adaptation of the Peat Accumulation Model (PAM; Hilbert et al. [2000]) made by Wu ([2012]). Wu ([2012]) modified PAM to study the response of peatland development and carbon cycling to climate change and to answer several research questions, among which was the following: How does peat accumulation respond to changes in precipitation and temperature? In its most basic form, PAM considers peat accumulation (i.e. SOL thickness) as a simple equation: "peat production minus the sum of oxic decomposition and anoxic decomposition equals peat accumulation" (i.e. change in SOL thickness = peat production – peat decomposition). In Wu's ([2012]) adaptation of PAM, the peat accumulation rate in ombrotrophic bogs could decrease by up to 70% within the first 100 years following the initiation of climate change. In fact, Wu's ([2012]) estimation suggested that for the 30 year period between 2011 and 2040, which corresponds to a 1°C increase in temperature, peat accumulation rate would drop by 15%. For the 30 year period between 2041 and 2070 (2°C increase in temperature) peat accumulation rate would drop by an additional 25%. Finally, for the 30 year period between 2071 and 2100 (3°C increase in temperature) peat accumulation rate would drop by an additional 30%. Hence, according to our estimations based on Wu's ([2012]) relationship between temperature increase and decrease in peat accumulation rate, this relationship fits the following equation:

$$\text{Peat}_{AR} = \left(16.0414 \times T\right) + \left(-5.7383 \times T^2\right) + \left(2.7230 \times T^3\right)$$

$$(3)$$

where Peat_{AR} is the accumulation rate of peat in $cm \cdot yr^{-1}$ and T the temperature (°C) increase. As a result, the peat accumulation rate decreases rapidly to the point where the accumulation rate was negative (i.e. peat decomposition rate was greater than production rate). Consequently, a temperature increase between 3°C and 4°C results in a reduction of SOL thickness of approximately 0.5 mm · yr^{-1}, between 4°C and 5°C a reduction of 1.8 mm · yr^{-1}, and between 5°C and 6°C a reduction of 3.8 mm · yr^{-1}. Hence, when a 4°C increase is reached, peat decomposition rate is greater than its production rate, leading to a reduction in SOL thickness. Although we recognize that this extension is simplistic and should be interpreted with greater caution, we believe this extension should be representative of the effect of temperatures increases beyond 3°C on peat accumulation rate and its effect on SOL thickness. Furthermore, although we acknowledge that ombrotrophic bogs and paludified forest support a different aboveground vegetation structure (i.e. shrubs- vs. tree-dominated aboveground vegetation for ombrotrophic bogs and paludified forests, respectively), we assumed that peat decomposition in ombrotrophic bogs and paludified forests would show similar responses to temperature increase.

From this information, we projected peat accumulation and SOL thickness according to three temperature increase scenarios (i.e. +1°C, +3°C and +6°C, which respectively correspond to 5%, 40% and 70% reduction in peat accumulation rate) and for four initial SOL thickness (5 cm, 10 cm, 20 cm, and 40 cm). For each scenario, average temperature was progressively increased during the simulations so that by 2100 temperature increases amount to +1°C, +3°C and +6°C relative to the average temperature as of 2004. For these four initial SOL thicknesses, peat accumulation rate varied according to Lecomte et al. ([2006]), i.e. where SOL is < 20 cm thick, peat accumulated at a rate of 10 cm per century, whereas where SOL is > 20 cm thick, peat accumulated at a rate of 20 cm per century. At each time step, SOL thickness was estimated as follow:

$$SOL_t = SOL_i + (30 \times Peat_{ARt})$$

(4)

where SOL_i is the thickness of the soil organic layer at the beginning of the reference period and $Peat_{ARt}$ is the accumulation rate of peat for the reference period. $Peat_{ARt}$ was calculated as follow:

$$Peat_{ARt} = Peat_{ARi} - (Peat_{ARi} \times (R_t/100))$$

(5)

where $Peat_{ARi}$ is the initial accumulation rate of peat and R_t is the reduction of peat accumulation rate for the period of reference, i.e. accumulation rate for each 30 year period. For each SOL thickness class (i.e. 5 cm, 10 cm, 20, and 40 cm), we calculated peat accumulation rate and the SOL thickness for each time step under the three climatic scenarios (i.e. +1°C, +3°C and +6°C).

Finally, in the index considering both climate change and natural fire regime, we first projected forest paludification considering the current fire cycle of 400 years as determined by Bergeron et al. ([2004]). Then we projected forest paludification allowing the fire cycle to decrease according to projections made by Bergeron et al. ([2010]). In this subindex, the fire cycle decreased to 300 years for 2011–2041, to 250 years for 2041–2071, and to 200 years for 2071–2100. Furthermore, based on our own observations, we considered that currently 50% of fire ignitions resulted in high-severity fires, i.e. fires that left < 5 cm of residual SOL over the mineral soil (Simard et al. [2007]). In light of greater uncertainty surrounding the impact of the climate change on fire severity, we ran our projections considering three proportions of high-severity fires, i.e. 25%, 50% and 75%. Furthermore, we considered that stands with SOL > 120 cm could not be submitted to high-severity fires because it is highly unlikely that a fire would leave < 5 cm of residual SOL over the mineral soil in such stands. At each time step, fire events and severity were randomly attributed to the forest stands of the study area.

The three indices were run in the following sequence: (1) base index, (2) base index + climate change, and (3) base index + climate change + natural fire regime modifications. This sequence allowed us to first explore the effect of climate change alone and then the combined effects of climate change and modifications of the fire regime on the potential for paludification of the forest stands of the study area.

RESULTS

Base Paludification Index

According to the Ontario forest map data, 42.2% of the area of the Gordon Cosens Forest was already *paludified* in 2004, whereas 57.8% of the area was *not paludified* (Table 3).

Table 3: Paludification level (% of the territory) of the Gordon Cosens forest according to the base index and the combination of the base index with climate change

Paludification level		Base index			Base index + climate change		
	Current[1]	2041	2071	2100	2041	2071	2100
Not paludified	57.8	57.8	46.1	31.4	57.8	57.8	57.8
Nearly paludified	0.0	0.0	11.7	26.4	0.0	0.0	0.0
Paludified	42.2	42.2	42.2	42.2	42.2	42.2	42.2

[1]Based on the Ontario's 2004 Forest Resource Inventory.

Lafleur et al.

Lafleur et al. Forest Ecosystems 2015 2:3, doi:10.1186/s40663-015-0027-6

As paludification is a relatively slow process, the index did not forecast any change in the level of paludification of the territory for the 2041 time step compared to the current state of the Gordon Cosens Forest (Table 3). As an illustration, Figure 3 shows the slow but constant thickening of the SOL for four different initial SOL thickness scenarios. When initial SOL thickness was set at 5 cm or 10 cm, SOL thickness projected for 2100 remained below 20 cm (Figure 3). However, when initial SOL thickness was set at 20 cm, SOL thickness reached 30 cm around 2060 and nearly 40 cm in 2100 (Figure 2). Similarly, when initial SOL thickness was set at 40 cm, SOL thickness reached 50 cm around 2060 and nearly 60 cm in 2100 (Figure 3). As a result, for the 2071 time step, the proportion of the study area occupied by *nearly paludified* stands increased from 0 to 11.7%, whereas as that occupied by *not paludified* stands decreased to 46.1% (Table 3); the proportion of the study area classified as *paludified* did not change. For the 2100 time step, the index forecasted that the proportion of the study area classified as *nearly paludified* increased to 26.4% (Table 3), while the proportion of area classified as *not paludified* decreased to 31.4%. At this time step, nearly 70% of the forest stands of the Gordon Cosens Forest could be classified as *paludified* or *nearly paludified*, which corresponds to a 166% increase in the cover of *paludified* or *nearly paludified* areas.

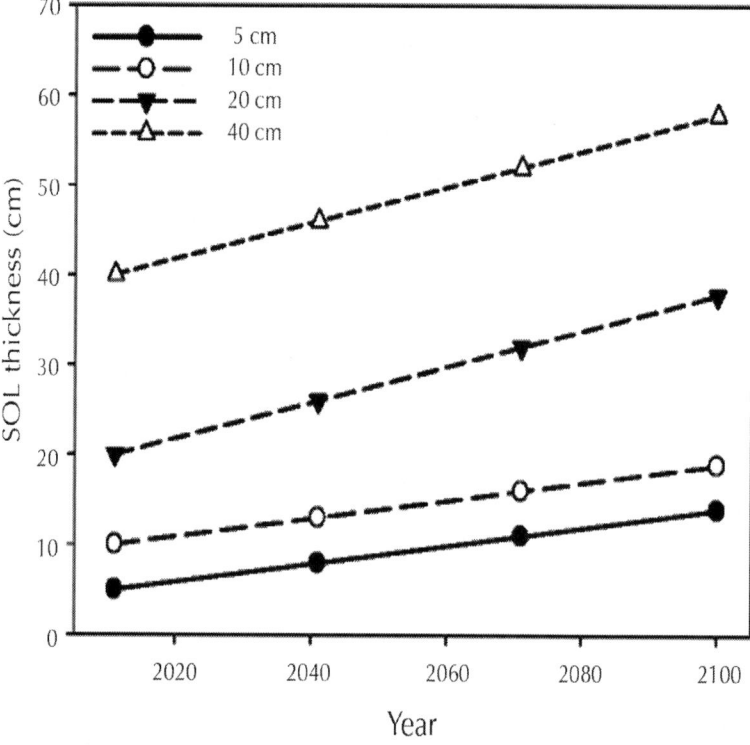

Figure 3: Soil organic layer (SOL) thickening with time without considering the effects of climate change. Initial SOL thickness (cm) is illustrated for four different scenarios: 5 cm, 10 cm, 20 cm, and 40 cm.

SOL Thickness and Climate Change

For every initial SOL thickness (i.e. 5 cm, 10 cm, 20 cm and 40 cm), SOL thickness tended to increase with time for the +1°C and +3°C scenarios (Figure 4), the increase being greater for the +1°C scenario. For the +6°C scenario, an increase was observed until ca. 2060 when SOL thickness started to decrease. This tipping point corresponds to the moment where the temperature increase was ca. 4°C. For the stands where initial SOL thickness was 5 cm or 10 cm, the +6°C scenario led to an almost complete disappearance of the SOL by 2100. For the stands where initial SOL thickness was 20 cm or 40 cm, the same scenario ended with SOL thickness similar to what was observed initially.

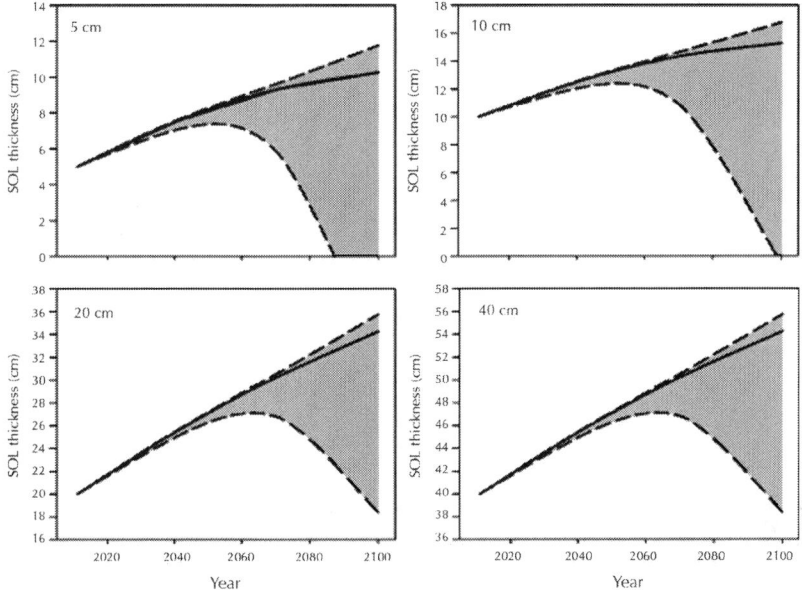

Figure 4: Soil organic layer (SOL) thickness (cm) variation with time in response to climate change. Each panel represents a different initial SOL thickness (i.e. 5 cm, 10 cm, 20 cm, and 40 cm) at the start of the projection. For each panel, upper dashed line = +1°C scenario; middle solid line = + 3°C scenario; lower dashed line = + 6°C scenario.

These negatives effect of climate change on peat accumulation resulted in no change in the proportion of the study area classified as *paludified, nearly paludified* and *not paludified* between 2011 and 2100 (Table 3).

Combining Climate Change with Natural Fire Regime

The inclusion of the current fire cycle (400 years) in our index resulted in a slight decrease (<6%) in the paludified area, regardless of the proportion of burnt stands submitted to high-severity fires (Figure 5a). Nonetheless, the decrease in paludified area was lower (ca. 2%) when 25% of the burned stands were submitted to high-severity fires and steeper (ca. 6%) when 75% were submitted to high-severity fires.

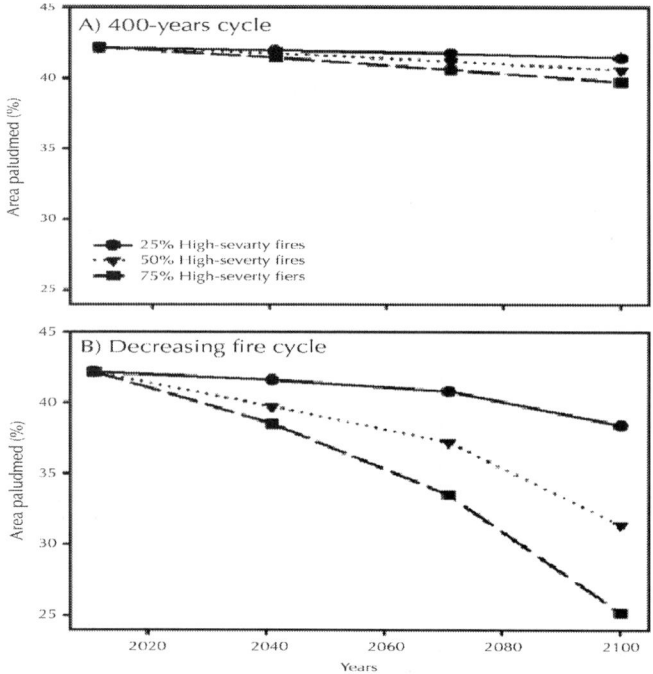

Figure 5: Projected effects of climate change and fire severity on paludified area for the period 2011–2100. A) Fire cycle was maintained at 400 years throughout the projection. B) Fire cycle was adjusted according to projections by Bergeron et al. ([2010]) for a region covering northeastern Ontario and northwestern Quebec; in 2041, fire cycle = 300 years; in 2071, fire cycle = 250 years; in 2100, fire cycle = 200 years.

When the fire cycle was allowed to decrease according to the projections made by Bergeron et al. ([2010]), declines in paludified area was steeper for the three proportions of high-severity fires (Figure 4b). Declines in paludified area were ca. 9%, 25% and 40% for the 25%, 50% and 75% high-severity fires, respectively (Figure 5b).

DISCUSSION

The boreal forest represents one of the Earth's largest biome, encompassing an area of approximately 14×10^6 km^2 (Wieder et al. [2006]). According to Bhatti et al. ([2003]), it is also one of the Earth's biomes most affected by global warming. About 25% of the boreal forest region is occupied by peatlands (Wieder et al. [2006]), a large proportion of which are forested peatlands. Overall, peatlands store approximately 15% of the Earth's terrestrial carbon (Gorham [1991]; Lavoie et al. [2005]). Consequently, at the global scale, any modifications to the natural fire regime in response to climate change may have a significant impact on the carbon sequestered in paludified forests. Our results suggest that in this context paludified forests may turn from C sink to atmospheric C source (in turn increasing atmospheric CO_2 concentration and providing a positive feedback on climate warming) if the increase in forest productivity does not compensate for carbon losses. However, it is important to note that the thick *Sphagnum* layer typical of paludified forests may protect SOL from drought and potential increasing depth of burn (and hence CO_2 emissions), and therefore that a significant proportion of the territory has high potential to remain paludified (Magnan et al. [2012]; Terrier et al. [2014a]).

The base paludification index produced slow but noticeable increases in proportion of the study area classified as *paludified* or *nearly paludified*. In the absence of fire in 2100 nearly 70% of the area of the Gordon Cosens Forest could be classified as nearly paludified (26%) or paludified (42%). This change is attributable to increased SOL thickness and concomitant change in moisture regime. Stands most susceptible to paludification were generally located on lacustrine surficial deposits or clay till, had fine textured soil (clay to coarse loam), a moisture regime classified as *Moist* to *Very moist*, SOL median depth of 20 cm, and an overstorey comprised of black spruce (sometimes accompanied by trembling aspen) (data not shown). Over the projected time frame considered with this base index, most of the stands of the study area were classified as *nearly paludified* or *paludified*.

Furthermore, as expected, temperature increase and the concomitant reduction in peat accumulation rate had an effect on SOL thickness. For the +1°C and +3°C scenarios, SOL thickness increased with time, although at a slower rate than in the model not considering climate change. For the +6°C scenario, however, peat accumulation ceased and SOL thickness started to decrease around the year 2060, indicating that the peat production rate was subsequently lower than its decomposition rate. In a long-term simulation, Ise et al. ([2008]) observed that a 4°C air temperature rise caused a 15% loss of soil organic carbon (measured in kg $C \cdot m^{-2}$) for a reference period of 100 years, which corresponds to a 10 cm reduction in SOL thickness. This is similar to the difference we observed (ca. 15 cm) between the +1°C and the +6°C scenarios. These results suggest that paludified stands (i.e. stands where SOL thickness > 40 cm) will remain paludified in 2100.

In consequence, the introduction of the effects of temperature increase in the paludification index did not produce any modifications in the proportion of the study area whether classified as *paludified,nearly paludified* or *not paludified* compared to the base index. Hence, despite increasing temperature between 2011 and 2100, the level of paludification of the study area did not change over time. Although our model did not consider the effects of changes in precipitation on paludification, a recent modeling study suggested that in the boreal forest of eastern Canada precipitation could increase by 10%–20% by the end of the 21st century (McKenney et al. [2010]). An increase in precipitation would logically lead to an increase in peat accumulation rate via the influence of higher water tables and a subsequent decrease in peat decomposition rate (Silvola et al. [1996]; Ise et al. [2008]). However, increased temperature is expected to lead to an increase in evapotranspiration, therefore leaving less moisture in the system (Soja et al. [2007]; Wu [2012]) and increasing risks of drought. A decrease in moisture availability in the system would, in turn, induce a lowering of the water table, a thickening of the oxic layer, and an increase in substrate temperature. Together these modifications would depress the peat accumulation rate, but more importantly, increase the peat decomposition rate, resulting in a steady SOL thickness over time. Therefore as two factors are acting in opposite directions (i.e. peat accumulation with time vs. increase in peat decomposition rate in response to climate change), the expected effects of climate change (not considering modifications to the current fire cycle) in our study area could result in a territory similar to the present in terms of paludified area.

Introducing wildfire in the index produced a quite different picture of the study area. Keeping the fire cycle constant and at the current level (i.e. 400 years) produced only a slight decrease in paludified area. However, allowing the fire cycle to decrease with time (as projected by Bergeron et al. [2010]) reduced the paludified area by between 10% and 40% under the high-fire severity scenario. Decreased fire cycle and reduced paludified area could be attributed to the fact that the projected increase in precipitation (McKenney et al. [2010])

may not fully compensate for increase in temperature, thus creating conditions that are more prone to fire occurrence (Terrier et al. [2013]). Yet, we feel that an increase in the proportion of high-severity fires in areas already paludified is unlikely as a recent study conducted in the forested peatlands of the Clay Belt during an extreme drought year failed to detect any effect of drought on soil moisture (measured as gravimetric water content [%]; Terrier et al. [2014a]) where thick SOL (>40 cm) occurs. In the same study, the authors projected for the period 2071–2100 the effects of extreme drought on potential SOL depth of burn, and concluded that increase in drought conditions should not be sufficient to greatly modify SOL depth of burn in areas where thick SOL prevail with potential depth of burn up to 0.7 cm (Terrier et al. [2014a]) for any individual fire event occurring between early spring and late fall. This resistance of SOL to drought and burn is related to the presence of *Sphagnum* species which hyaline cells stock large amounts of water (Silvola [1991]) and limit its evaporation (Busby and Whitfield [1978]), and therefore reduce potential depth of burn. However, Terrier's study (Terrier et al. [2014a]) also suggested that in areas where SOL is < 40 cm and where *Pleurozium schreberi* is the dominant ground-covering moss, drought was able to depress soil moisture to the extent where they projected a potential depth of burn up to 3.2 cm for the period 2071–2100. These results suggest that areas that are in the process of being paludified exhibit a relatively high potential to remain unpaludified, and that those areas that are already paludified areas show a low potential for depaludification. These results are also in accordance with Magnan et al. ([2012]) who found no major changes in boreal peatlands despite evidence of slowed peat accumulation rates due to fire, and with Magnan et al. ([2014]) who showed that *Sphagnum*-dominated bogs located in a maritime environment have persisted over millennia and that fires had few impacts on their vegetation dynamics.

Thus, at the landscape level, when fire cycle is kept constant at the current level, the slow rate of paludification results in a slight decrease in paludified area regardless of the proportion of high severity fires, suggesting either that the rate of peat accumulation in the index is too conservative, or that the current highly paludified landscape is not completely in balance with the current climate and fire regime (Payette [2001]). Similarly, a decreasing fire cycle is likely to limit the development of paludified stands.

CONCLUSIONS

Overall, our models predict a slight (no modification to the current fire cycle) to moderate (decreasing fire cycle over time) decrease in area covered by paludified forests within the Gordon Cosens Forest in 2100. This might have several impacts for the forest industries of the Clay Belt as slower paludification rates imply greater forest productivity and potentially a greater potential for forest harvest.

Furthermore, at the global scale, increasing fire frequency in boreal paludified forests may have important consequences on carbon storage and climate if increases in forest productivity do not compensate for carbon losses. However, as the thick *Sphagnum* layer typical of paludified forests may protect soil organic layer from drought and deep burns, a significant proportion of the territory has high potential to remain a carbon sink.

In this context, our findings are supported by those of Terrier et al. ([2014b]) who in an area adjacent to our study area modeled the impacts of climate change on fire regime, vegetation dynamics and SOL depth of burn. They concluded that although climate change is likely to increase burn rates, the moist and cool conditions in these forests would prevent high depth of burn and the landscape would remain paludified.

This simple tool could be used by forest managers to forecast the development of paludified forests and to plan forest operations and conservation areas, and by policy makers to plan carbon management at the regional scale. Nonetheless, we recognize that in order to strengthen the predictions made by our index, the next step would be to construct a mechanistic model that includes sensitivity analysis as well as fire sub-indices that take into account drought and wildfire risk.

AUTHORS' CONTRIBUTIONS

BL developed the models, carried out the analysis, and wrote the manuscript, NJF and YB were instrumental in conceptualizing and refining the models, and in revising the manuscript. All authors read and approved the final manuscript.

ACKNOWLEDGEMENTS

This study was made possible by funding from the Ontario Ministry of Natural Resource. We thank Rachelle Lalonde for providing forest maps, Mélanie Desrochers for map production, and Aurélie Terrier for helpful discussion on fire modeling.

REFERENCES

1. Bergeron Y, Gauthier S, Flannigan M, Kafka V: Fire regimes at the transition between mixedwood and coniferous boreal forest in northwestern Quebec. *Ecology* 2004, 85:1916-1932.

2. Bergeron Y, Cyr D, Girardin MP, Carcaillet C: Will climate change drive 21st century burn rates in Canadian boreal forest outside of its natural variability: collating global climate model experiments with sedimentary charcoal data. *Int J Wildl Fire* 2010, 19:1127-1139.

3. Bhatti JS, Van Kooten GC, Apps MJ, Laird LD, Campbell ID, Campbell C, Turetsky MR, Yu Z, Banfield E: Forest management planning based on natural disturbance and forest dynamics. In *Towards sustainable management of the boreal forest*. Edited by Burton PJ, Messier C, Smith DW, Adamowicz WL. NRC Research Press, Ottawa; 2003:799-855.

4. Busby JR, Whitfield DWA: Water potential, water content, and net assimilation of some boreal forest mosses. *Can J Bot* 1978, 56:1551-1558.

5. de Groot WJ, Pritchard JM, Lynham TJ: Forest floor fuel consumption and carbon emissions in Canadian boreal forest fires. *Can J For Res* 2009, 39:367-382.

6. Drobyshev I, Simard M, Bergeron Y, Hofgaard A: Does soil organic layer thickness affect climate-growth relationships in the black spruce boreal ecosystem? *Ecosystems* 2010, 13:556-574.

7. Dyrness CT, Norum RA: The effects of experimental fires on black spruce forest floors in interior Alaska. *Can J For Res* 1983, 13:879-893.

8. Environment Canada (2014) Canadian climate normals 1981–2010. http://www.climate.weather.gc.ca/climate_normals/index_e.html. Accessed 20 Oct 2014

9. Flannigan M, Logan KA, Amiro B, Skinner W, Stocks BJ: Future area burned in Canada. *Clim Change* 2005, 72:1-16.

10. Girardin MP, Mudelsee M: Past and future changes in Canadian boreal wildfire activity. *Ecol Appl* 2008, 18:391-406.

11. Gorham E: Northern peatlands: role in the carbon cycle and probable responses to climatic warming. *Ecol Appl* 1991, 1:182-195.

12. Greene DF, Macdonald SE, Cumming S, Swift L: Seedbed variation from the interior through the edge of a large wildfire in Alberta. *Can J For Res* 2005, 35:1640-1647.

13. Hilbert DW, Roulet N, Moore T: Modelling and analysis of peatlands as dynamical systems. *J Ecol* 2000, 88:230-242.

14. IPCC: Stocker TF, Qin D, Plattner G-K, Tignor M, Allen SK, Bashung J, Nauels A, Xia Y, Bex V, Midgley PM (Eds): Climate change 2013: The physical science basis In *Contribution of Working Group I to the Fifth Assessment Report of the Intergovernmental Panel on Climate Change*. Cambridge University Press, Cambridge, UK; 2013.

15. Ise T, Dunn AL, Wofsy SC, Moorcroft PR: High sensitivity of peat decomposition to climate change through water-table feedback. *Nat Geosci* 2008, 1:763-766.

16. Jeglum JK: Definition of trophic classes in wooded peatlands by means of vegetation types and plant indicators. *Ann Bot Fenn* 1991, 28:175-192.

17. Lafleur B, Fenton NJ, Paré D, Simard M, Bergeron Y: Contrasting effects of season and method of harvest on soil properties and the growth of black spruce regeneration in the boreal forested peatlands of eastern Canada. *Silva Fenn* 2010, 45:799-813.

18. Lafleur B, Paré D, Fenton NJ, Bergeron Y: Do harvest and soil type impact the regeneration and growth of black spruce stands in northwestern Quebec? *Can J For Res* 2010, 40:1843-1851.

19. Lavoie M, Paré D, Bergeron Y: Impact of global change and forest management on carbon sequestration on northern forested peatlands. *Environ Rev* 2005, 13:199-240.

20. Lecomte N, Simard M, Bergeron Y, Larouche A, Asnong H, Richard PJH: Effects of fire severity and initial tree composition on understorey vegetation dynamics in a boreal landscape inferred from chronosequence and paleoecological data. *J Veg Sci* 2005, 16:665-674.

21. Lecomte N, Simard M, Fenton N, Bergeron Y: Fire severity and long-term ecosystem biomass dynamics in coniferous boral forests of eastern Canada. *Ecosystems* 2006, 9:1215-1230.

22. Magnan G, Lavoie M, Payette S: Impact of long-term vegetation dynamics of ombrotrophic peatlands in northwestern Québec, Canada. *Quat Res* 2012, 77:110-121.

23. Magnan G, Garneau M, Payette S: Holocene development of maritime ombrotrophic peatlands of the St. Lawrence North Shore in eastern Canada. *Quat Res* 2014, 82:96-106.

24. McKenney DW, Pedlar JH, Lawrence K, Gray PA, Colombo SJ, Crins WJ: *Current and projected future climatic conditions for ecoregions and selected natural heritage areas in Ontario.* Ontario Forest Research Institute, Ministry of Natural Resources, Sault Ste. Marie, ON, Canada; 2010.

25. Payette S: Les principaux types de tourbières. In *Écologie des tourbières du Québec-Labrador.* Edited by Payette S, Rochefort L. Les Presses de l'Université Laval, Québec, Québec; 2001:39-89.

26. Payette S, Garneau M, Delwaide A, Schaffhauser A: Forest soil paludification and mid-Holocene retreat of jack pine in easternmost North America: Evidence for a climatic shift from fire-prone to peat-prone conditions. *The Holocene* 2013, 23:494-503.

27. Riley JL: *Peat and peatland resources of northeastern Ontario.* Ministry of Northern Development and Mines, Ontario Geological Survey. Misc. Paper No; 1994.

28. Silvola J: Moisture dependence of CO_2 exchange and its recovery after drying in certain boreal forest and peat mosses. *Lindergia* 1991, 17:5-10.

29. Silvola J, Alm J, Ahlholm U, Nykanen H, Martikainen PJ: CO_2-fluxes from peat in boreal mires under varying temperature and moisture conditions. *J Ecol* 1996, 84:219-228.

30. Simard M, Lecomte N, Bergeron Y, Bernier PY, Paré D: Forest productivity decline caused by successional paludification of boreal soils. *Ecol Appl* 2007, 17:1619-1637.

31. Simard M, Bernier PY, Bergeron Y, Paré D, Guérine L: Paludification dynamics in the boreal forest of the James Bay Lowlands: effect of time since fire and topography. *Can J For Res* 2009, 39:546-552.

32. Sims RA, Baldwin KA: *Forest humus forms in northwestern Ontario. Natural Resources Canada, Canadian Forest Service, Great Lakes Forestry Centre, Technical Report TR-28*. 1996.

33. Sims RA, Towill WD, Baldwin KA, Wickware GM: *Field guide to the forest ecosystem classification for northwestern Ontario*. 1989.

34. Soil Classification Working Group (1998) The canadian system of soil classification, 3rd edn. Agriculture and Agri-Food Canada Publication 1646, Canada

35. Soja AJ, Tchebakova NM, French NHF, Flannigan MD, Shugart HH, Stocks BJ, Sukhinin AI, Parfenova EI, Chapin FS III, Stackhouse PW: Climate-induced boreal forest change: Predictions versus current observations. *Global Planet Change* 2007, 56:274-296.

36. Taylor KC, Arnup RW, Merchant BG, Parton WJ, Nieppola J: *A field guide to forest ecosystems of Northeastern Ontario*. 2nd edition. Queen's Printer for Ontario, Canada; 2000.

37. Terrier A, Girardin MP, Périé C, Legendre P, Bergeron Y: Potential changes in forest composition could reduce impacts of climate change on boreal wildfires. *Ecol Appl* 2013, 23:21-35.

38. Terrier A, de Groot WJ, Girardin MP, Bergeron Y: Dynamics of moisture content in spruce-feather moss and spruce-*Sphagnum* organic layers during an extreme fire season and implications for future depths of burn in Clay Belt black spruce forests. *Int J Wildl Fire* 2014, 23:490-502.

39. Terrier A, Girardin MP, Cantin A, de Groot WJ, Anyomi KA, Gauthier S, Bergeron Y (2014b) Disturbance legacies and paludification mediate the ecological impact of an intensifying wildfire regime in the Clay Belt boreal forest of eastern North America. J Veg Sci doi:10.1111/jvs.12250.

40. van Bellen S, Garneau M, Bergeron Y: Impact of climate change on forest fire severity and and consequences for carbon stocks

in boreal forest stands of Quebec, Canada: A synthesis. *Fire Ecol* 2010, 6:16-44.

41. Veillette JJ: Evolution and paleohydrology of glacial lakes Barlow and Ojibway. *Quat Sci Rev* 1994, 13:945-971.

42. Wieder RK, Vitt DH, Benscoter BW: Peatlands and the boreal forest. In *Boreal peatland ecosystems*. Edited by Wieder RK, Vitt DH. Springer, New York, NY; 2006:1-8.

43. Wu J: Response of peatland development and carbon cycling to climate change: a dynamic system modeling approach.*Environ Earth Sci* 2012, 65:141-151.

44. Yu Z, Campbell ID, Vitt DH, Apps MJ: Modelling long-term peatland dynamics. I. Concepts, review, and proposed design. *Ecol Model* 2001, 145:197-210.

Priority Areas for Watershed Service Conservation in the Guapi-Macacu Region of Rio de Janeiro, Atlantic Forest, Brazil

Vanesa Rodríguez Osuna[1, 2], Jan Börner[1], Udo Nehren[2], Rachel Bardy Prado[3], Hartmut Gaese[1], and Jürgen Heinrich[4]

[1]Center for Development Research (ZEF), Department of Economic and Technological Change, University of Bonn, Walter-Flex-Str. 3, Bonn, 53117, Germany

[2]Institute for Technology and Resources Management in the Tropics and Subtropics (ITT), Cologne University of Applied Sciences, Betzdorfer Str. 2, Cologne, 50679, Germany

[3]Empresa Brasileira de Pesquisa Agropecuária (EMBRAPA Soils), Brazilian Enterprise for Agricultural Research, Geotechnologies and Environmental Monitoring, Rua Jardim Botânico, Rio de Janeiro, CEP 22460-000, Brazil

[4]Institute for Geography, Department of Physical Geography, University of Leipzig, Johannisallee 19a, Leipzig, 04103, Germany

ABSTRACT

Introduction

Land use intensification and urbanisation processes are degrading hydrological ecosystem services in the Guapi-Macacu watershed of Rio de Janeiro. A proposal to pay farmers to restore natural watershed services might be an alternative to securing the water supply in the long-term for the around 2.5 million urban water users in the study region. This study quantifies the costs of changing current land use patterns to enhance watershed services and compares these costs to the avoided costs associated with water treatment for public supply.

Methods

We use farm-household data to estimate the opportunity costs of abandoning current land uses for the recovery of natural vegetation; a process that is very likely to improve water quality in terms of turbidity due to reduced inputs from erosion. Opportunity cost estimates are extrapolated to the watershed scale based on remote sensing land use classifications and vulnerability analysis to identify priority zones for watershed management interventions. To assess the potential demand for watershed services, we analyse water quality and treatment cost data from the main local water treatment plant.

Results

Changing agricultural land uses for watershed services provision generally comes at high opportunity costs in our study area near to the metropolis of Rio de Janeiro. Alternative low cost watershed conservation options do exist in the livestock production sector. These options have the potential to directly reduce the amount of sediments and nutrients reaching the water bodies, and in turn decrease the costs of treatment needed for drinking water. Land cover changes at the scale needed to improve water quality will, nonetheless, likely exceed the cost of additional investments in water treatment.

Conclusions

The state water utility company's willingness to pay for watershed services alone will not be enough to induce provision of additional watershed services. We conclude that monetary incentives conditioned on specific adjustments to existing production systems could still have a complementary role to play in improving watershed services. However, we note that our willingness to pay analysis focusses on only one of the potentially wide range of ecosystem services provided by natural vegetation in the Guapi-Macacu watershed. Factoring these ecosystem services into the willingness to pay equation is likely to change our assessment in favour of additional conservation action, be it through PES or other policy instruments.

INTRODUCTION

One of the biggest challenges of our time is to revert the ongoing degradation of ecosystems while meeting the increasing demand for food and biomass (Thomas and Callan [2010]; MEA [2005]). Population and economic growth are increasing the demand for water resources and, at the same time, amplifying the pressure on ecosystems that deliver watershed services (Porras et al. [2008]).

The degradation of ecosystem services represents a loss of widely undervalued natural capital assets (TEEB [2009]; Montes [2007]; MEA [2005]. While there is general agreement that land use choices influence watershed service provision, the magnitude and nature of the effects are highly context dependent and poorly understood in tropical forest environments (Porras et al. [2008]; Veiga [2008]; Calder [2005]; van Noordwijk [2005]).

Land and water linkages are challenging to manage as watershed services have a common good characteristic (Porras et al. [2008]) and are generally unaccounted for in monetary terms. As a result, they are often degraded and this is not reflected in traditional economic measures, such as GDP (TEEB [2009]; Costanza et al. [1997]; FAO [2007]). Land use patterns and land management practices by upstream landholders in a watershed determine, to a great extent, quality and quantity of water bodies (Porras et al. [2008]). For example, unsustainable land use and agricultural practices can result in negative hydrological side-effects, or externalities, such as sedimentation (Veiga [2008]; MEA [2005]). As a result, the design of policy measures that encourage farmers to adopt watershed conservation measures and, at the same time, maintain or even increase productivity has become a major research issue.

Several policy options are available to enhance incentives for the supply of ecosystem services. Command-and-control regulations, such as bans on forest clearing, have long been the preferred policy option to control environmentally harmful land use changes (Boerner and Vosti [2012]; Porras et al. [2008]; FAO [2007]). More recently, mechanisms based on economic incentives or markets for ecosystem services are being proposed as cost-effective complementary tools to promote ecosystem service conservation (Porras et al. [2008]; Engel et al. [2008]; FAO [2007]). These incentive-based mechanisms include payments for ecosystem services (PES) and payments for watershed services (PWS), water quality trading markets and reciprocal or in-kind agreements (Bennett et al. [2013]).

Latin America registered a transaction value of USD 528.9 million in such investments between 2001 and 2011 corresponding to 3.4 million hectares (Bennett et al. [2013]). PWS programmes are considered to be the "fastest growing" and most mature among all PES schemes in Latin America (Bennett et al. [2013]; Balvanera et al. [2012]; Stanton et al. [2010]). PWS in Latin America typically involve compensating upstream rural producers for protecting and/or restoring natural forest ecosystems and highland natural pastures (páramo) (Pria et al. [2013]; Veiga and Galvadao [2011]; Grieg-Gran et al. [2005]). Such ecosystems are mostly located in strategic water production areas, such as headwaters, riparian forest or water intake points for water supply (Pria et al. [2013]; Veiga and Galvadao [2011]). There is, however, little evidence on the effectiveness of such schemes in meeting conservation and development goals in the peer-reviewed literature (see for example Arriagada et al. [2012]; Pattanayak et al. [2010]).

In Brazil, PWS are expanding and there are already 848 ecosystem service providers in the Atlantic Forest region alone, mainly organised or supported within the National Water Agency's Water Producer Programme (Veiga and Galvadao [2011]; Santos et al. [2010]). As of 2012, 41 PWS projects had been implemented or were in preparation, covering an area of around 40,000 hectares. A project in the state of Rio de Janeiro is currently in development, in the Três Picos State Park, Municipality of Cachoeiras de Macacu (Guedes and Seehusen [2011]). This State Park is located in the mountainous area of the Guapi-Macacu watershed, which contributes to the supply of drinking water for almost 2.5 million inhabitants of five municipalities, including the city of Niteroi (Pedreira et al. [2009]) and this watershed is the focus of the present study. The main drivers of degradation of water resources in this watershed are urbanisation processes, intensive agriculture and conversion of riparian vegetation.

Managing watershed services requires solid knowledge about (1) the costs of providing an additional unit of water quality or quantity (supply side) and (2) water users or intermediaries' willingness to pay for an additional unit of water quality or quantity (demand side). The economic aspects of watershed services supply and demand are particularly poorly studied and many PWS initiatives begin operating without quantitative knowledge of such parameters (Martin-Ortega et al. [2012]). To address this knowledge gap, we focus on two research questions within our study area of the Guapi-Macacu watershed in the state of Rio de Janeiro:

- What are the costs associated with shifting land uses to foster improved and enhanced watershed services (services supply)?
- What are the treatment costs for the water supply company to improve water quality parameters such as turbidity?

Our focus is on understanding the supply and environmental conditions that determine watershed service provision. Particularly, we concentrate on the watershed service related to water quality maintenance in terms of turbidity reduction for supplying drinking water. However, we also analyse factors that influence water treatment costs (related to services demand) based on land use changes. The remainder of this paper is structured as follows, section "Assessing watershed services supply and demand" presents our approach, addresses relevant literature and concepts related to watershed service supply and demand, section "Methods" describes the study area and methods used for this research, section "Results" presents our findings, section "Discussion" discusses these findings and contextualises them with other studies and reviews the main caveats. Finally, section "Conclusions" presents our main conclusions and the policy implications for watershed service management.

ASSESSING WATERSHED SERVICES SUPPLY AND DEMAND

The Supply of Watershed Services

Watershed service provision is to a large extent determined by land use and land cover. For example, unsustainable land use is frequently linked to a high surface runoff and an elevated concentration of suspended and soluble loads in water bodies (Batchelor et al. [1998]). Changes in land cover, such as forest to agriculture conversion, tend to increase superficial runoff and sediment flux in rivers (FAO[2007]; MEA [2005]).

Healthy forest and wetland ecosystems are considered very effective at regulating water flow and improving water quality (Russi et al. [2013]; TEEB [2010a] [b]). Maintaining water quality includes the control of sediment, nutrient (in particular phosphorous and nitrogen) and chemical loads, as well as salinity (TEEB [2010b]). In addition, forest ecosystems can remove pathogenic microbes, sequester and convert inorganic ions and transform persistent organic pollutants (TEEB [2010b]).

Watersheds with an important extension of forest tend to offer better water quality than those subject to other land uses, such as agriculture, pasture, industry and urban infrastructure, since the latter are associated with higher discharges of diverse types of pollutants into soils and water. In this way, the presence of forest could substantially reduce the cost of water treatment for drinking water in most cases, thereby reducing the related costs for water supply (Medeiros et al. [2011]; Reis [2004]).

Cities such as Rio de Janeiro, Johannesburg, Tokyo, Melbourne, New York and Jakarta all depend on protected areas with forests to provide drinking water for their residents (Dudley and Stolton [2003]). Moreover, a third of the hundred largest cities worldwide take a significant proportion of their drinking water from protected forested areas (Dudley and Stolton [2003]).

Worldwide several examples show that well-managed forests, wetlands and protected areas are very likely to supply clean water at significantly lower costs than man-made replacements, such as water treatment plants (Bennett et al. [2013]; Hanson et al. [2011]; TEEB [2009]; Postel et al. [2005]). Some examples are included in Table 1. For example, in the United States of America, 27 water suppliers showed that in watersheds with at least 60% forest cover, treatment costs were 50 percent less than those with 30% forest cover (Postel et al. [2005]). Accordingly, several North American cities have decided to invest in watershed management to avoid the expense of water treatment plants.

Table 1: Examples of the economic values of watershed services

Action	Examples	Source
Payments to maintain water purification services in the Catskills watershed, New York avoid the costs of launching a filtration plant for water treatment	Payments for maintaining watershed purification represent 1 to 1.5 billion USD in comparison to the much higher estimated cost of a filtration plant (6–8 billion USD plus 300–500 million USD yearly for operating costs)	Hanson et al. [2011]; WRI [2011]; Pagiola et al. [2004]
The cost of removing nitrate contamination from water in Rochester, Minnesota	2.8-4.8 million USD per year	Hawkins [2003]
WTP for water quality improvement in the levels of iron, sulphate, hardness, and copper in South-western, Minnesota	2.4, 2.0, 6.6 and 2.6 million USD yearly respectively (via contingent valuation method)	Hawkins [2003]

Yearly revenues in public utility resulting from natural water supply in Mud Lake, Minnesota/ South Dakota	94 USD per acre	Hawkins [2003]
Annual WTP from residents living close to the Minnesota River for reducing phosphorous levels	14.1 USD via taxes or 19.6 USD via water bills for a 40% reduction in a nearby river (contingent valuation method)	Hawkins [2003]
Increased reservoir water quality and surface area in St. Louis, Missouri	Yearly net benefit of 25 million USD (travel cost method)	Hawkins [2003]
Value of water supply in Milesburg, Pennsylvania	Between 14 and 36 USD per household (avoided cost)	Hawkins [2003]
In Venezuela, a national protected area system avoids sedimentation	Without the provision of this service by the national protected area, unavoided sedimentation could reduce farmer's income by approximately 3.5 million USD yearly	Pabon-Zamora et al.[2008]
The current provision of ecosystem services related to existing stream vegetation along the Llobregat River, Spain	79,000 EUR per year savings in water treatment costs for the residents of Barcelona	Honey-Roses et al.[2013]
In New Zealand, the value of water provision from the Otago region coming from the Te Papanui Conservation Park	Now it is "free" but would cost 136 million NZDa to bring it in from somewhere else (total benefit of the service)	TEEB [2009]

[a]1 USD is equivalent to 1.30 NZD at current exchange rate (http://www.oanda.com/currency/converter/).

Rodríguez Osuna et al.

Rodríguez Osuna et al. Ecological Processes 2014 3:16   doi:10.1186/s13717-014-0016-7

In the literature regarding water quality, off-site effects of soil erosion are frequently referred to as sedimentation (Veiga [2008]; Holmes [1988]). A summary of economic activities most affected by loss of watershed services is presented in Table 2.

Table 2: Examples of economic activities most affected by the loss of watershed services

Economic activities and damage caused by the loss of watershed services	Examples of incurred costs/ damages due to the loss of watershed services	Source
Activities that depend on reservoirs	Capacity loss for energy generation Capacity loss of irrigated production Loss of benefits related to flood control Capacity loss of navigation channels Increase of suspended material/ siltation of water bodies resulting in a reduction of their storage capacity	Veiga [2008]; Reis[2004]
Drainage and maintenance operations	Costs related to (a) irrigation and drainage of channels, (b) hydroelectric plant reservoirs, (c) ports and (d) eutrophication (increase concentration of N and P in water bodies)	Veiga [2008]; Reis[2004]

Increased water treatment costs related to augmented turbidity	One example are watersheds in São Paulo with lower costs related to consumption of chemical products for the water treating process (less than 20 BRLa per 1,000 m3 of treated water) were those with a forest cover higher than 15%. The two water treatment units with less percentage of forest cover (Piracicaba and Atibaia rivers with less than 10% forest cover) show considerably higher specific costs of chemical products	Reis [2004]; Dearmont et al.[1998]; Holmes[1988], Moore and McCarl [1987]
Monetary damages related to sedimentation	Additional annual costs from 35 to 661 million USD for a US water treatment plant due to soil erosion. These values relate strongly to agricultural production causing sedimentation	López [1997]; Holmes[1988]
	Additional annual operational cost of 3.2 BRL per hectare and 0.10 BRL per ton of eroded soil for the water treatment of the river Corumbataí in São Paulo. These estimates were made assuming that the only externality caused by soil erosion were increased water treatment costs but not taking into account siltation of river banks, flooding of areas close to river banks and loss of navigation capacity	

a1 USD is equivalent to 2.38 BRL at current exchange rate (http://www.oanda.com/currency/converter/). This rate will be used onwards in this document when comparison among monetary units is required.

Rodríguez Osuna et al.

Rodríguez Osuna et al. Ecological Processes 2014 3:16   doi:10.1186/s13717-014-0016-7

Different authors have classified ecosystem services in distinct ways (Haines-Young and Potschin[2013] Daily and Matson [2008]; FAO [2007]; Farber et al. [2006]; MEA [2005]; Postel et al. [2005]; Hawkins [2003]; De Groot et al. [2002]; Costanza et al. [1997]). In the current study, we followed the TEEB ([2010b]) classification of watershed services, whereby maintenance of water quality for human consumption is considered a provisioning service. We use this terminology throughout this article.

It is often argued that a major challenge of market mechanisms relates to the "packaging" of ecosystem services into commodities that are tradable or subject to a contract (Porras et al. [2008]). However, most PWS schemes in developing countries are guided by a "land-based" approach, whereby suppliers are paid to improve their land management practices, which are in turn considered highly likely to result in watershed service provision, rather than being paid for the actual service delivery (Porras et al. [2008]). Regardless of the approach chosen, watershed services supply is inevitably linked to farmers' land use decisions (FAO [2007]). Consequently, watershed services supply analyses often require cost-benefit analyses of agricultural production systems.

The concept of opportunity costs (OCs) is most frequently used to express the costs of additional watershed service provision. In the context of watershed services, OCs represent any benefits foregone by a farmer upon converting from their current land use to an alternative form of land use that is more watershed service-friendly.

Ideally, PWS are designed in such a way that payments compensate for at least the OCs of additional service provision or the foregone benefits of the land use promoted in order to provide the service. The extent to which such payments are or are not appropriate depends on the alternative land uses in each given area (Pagiola et al. [2010]).

The Demand for Watershed Services

When dealing with demand, we refer to those who are currently benefiting from the delivery of watershed services and to the resources available for protecting and conserving these services (Guedes and Seehusen [2011]).

Besides PWS, which are carried out by national states in Latin America, a user's fee system can be an effective approach to efficiently use water resources. The watersheds of the rivers Piracicaba, Capivari and Jundiaí (PCJ) in the state of São Paulo have implemented such a user's fee system with around 8.8 million of beneficiaries of the Cantareira system (Veiga and Galvadao [2011]). In this situation, an Inter-Municipal Basin Committee was formed to manage a watershed protection fund and contributions to the fund come from the municipal water utility budgets.

As a further alternative, some programmes in the Atlantic Forest region are subsidised by the government, for example: "Bolsa Verde", "ProdutorES de Água" and "Mina D'água" in the states of Minas Gerais, Espírito Santo and São Paulo, which pay rural producers for conservation activities on their properties (Guedes and Seehusen [2011]). These include the protection or restoration of native vegetation areas with a focus on headwaters and riparian forests (Guedes and Seehusen [2011]; Veiga and Galvadao [2011]).

Another significant source of finance for PWS in developing countries comes from the international public sector funding or development assistance (Porras et al. [2008]). A key provider to this funding is the Global Environmental Facility (GEF), which acts as buyer on behalf of service users for conserving global public services. Around 108 million USD and 52 million USD have been made available as World Bank (WB) loans and GEF grants respectively for WB/GEF-supported PWS projects (FAO [2007]). For example, the World Bank has given loans to support the development and implementation of well-known PES programmes in Mexico and Costa Rica (FAO [2007]). Involvement of the private sector in paying for ecosystem services, including in the context of corporate social responsibility, is growing (TEEB [2010c]).

Considering these factors, supply and demand analysis of watershed services provides essential information to assess the economic and financial viability of PWS schemes (IIED [2012a]). Figure 1 illustrates our approach in identifying the economic preconditions for a potential PWS scheme in the Guapi-Macacu watershed.

WATERSHED SERVICE (WS): WATER QUALITY MAINTENANCE FOR DRINKING WATER PUBLIC SUPPLY

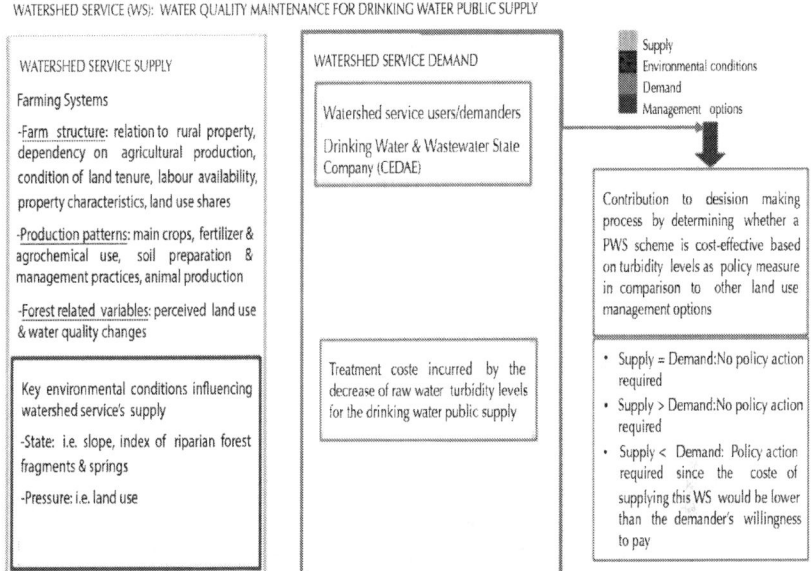

Figure 1: Conceptual framework for assessing watershed services. This framework considers water quality maintenance for public supply considering supply is at the farming system level and demand is at the watershed level.

As the figure suggests, we expect PWS to be a viable policy option only if the willingness of water users to pay (i.e. the demand of the water company) exceeds the opportunity costs (OCs) of additional watershed service provision (supply) incurred by land users in the watershed.

METHODS

The Guapi-Macacu Watershed

The study area lies within the Guapi-Macacu watershed (1,263 km²), which is located in the Serra do Mar biogeographical region (Ribeiro et al. [2009]) and is a priority conservation target within the Atlantic Forest biodiversity hotspot in the state of Rio de Janeiro (CEPF [2001]), see Figure 2.

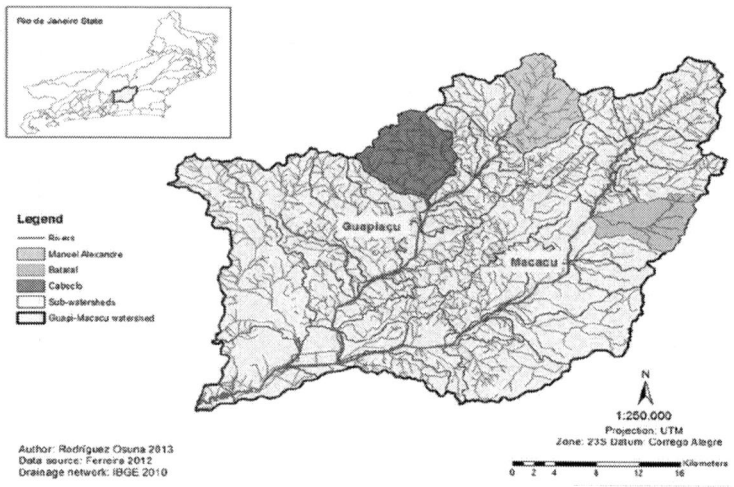

Figure 2: The Guapi-Macacu watershed within the state of Rio de Janeiro.

The Atlantic Forest biome is characterised by historically high deforestation rates (Dean [1997]), resulting in a highly fragmented area with numerous isolated and disperse forest fragments in a landscape dominated by agricultural production systems (Nehren et al. [2013]; Ribeiro et al. [2009]). This biome supplies 135 million people with water (Pria et al. [2013]); however, a mere 11 to 16 percent of its original forest cover is left (Ribeiro et al. [2009]).

Forest cover in the state of Rio de Janeiro and more specifically in the Guapi-Macacu watershed, is comparatively high due to its rugged topography. Unfortunately, continuous forest cover occurs mostly on steep slopes that are inappropriate for agriculture, while the foothills and lowlands are dominated by agriculture and pasture (Nehren et al. [2013]; Strobel et al. [2007]). In 2008, the land cover/land use of the Guapi-Macacu watershed consisted of forests in all stages (48.8%), pastures (41.4%) and agriculture (4.4%). The remaining 5.4% were covered by urban areas, water bodies, bare soil, rock outcrop, wetlands and mangrove (Fidalgo et al. [2008]).

Rivers and creeks in the Guapi-Macacu watershed originate mainly in the State Park "Três Picos", but some stem from the National Park "Serra dos Órgãos" and in the State Ecological Station "Paraíso". The main rivers Macacu and Guapiaçu originate within the State Park and constitute the main components of this watershed (Strobel et al. [2007])

The water intake point of the water supply is located in the lower part of the watershed and managed by the Drinking Water and Wastewater State Company (CEDAE). Due to the good water quality coming from the springs of the rivers Macacu and Guapiaçu, several mineral water companies have been established in the region, as well as enterprises for which water is an essential input (Strobel et al. [2007])

In the Guapi-Macacu watershed, various demands in water supply to domestic, industrial and agricultural consumption are taking place as a result of the good water quality from the water sources of the main rivers (Strobel et al. [2007]). Moreover, the ongoing construction of a new petrochemical complex (COMPERJ) in the downstream area of the watershed is expected to further increase both water demand and pressure on natural resources in the study area (Pedreira et al. [2009]).

Within the watershed, three sub-watersheds (see Figure 3): (I) Manuel Alexandre, (II) Batatal and (III) Caboclo were selected for this study. Each of these sub-watersheds represents the different types of land cover mix found in the area, namely "forest", "agriculture" and "pasture", which account altogether for around 95% of the total watershed area. By selecting three sub-watersheds with greatly varying land use patterns and intensities, we were able to compare them with respect to their differences on agricultural profitability and their distinctive impact on water quality parameters: in this case turbidity levels.

- The sub-watershed Manuel Alexandre is located in the Ecological Reserve of Guapiaçu (REGUA) and represents a well-preserved landscape with a high proportion of forest land (87%) (Fidalgo et al.[2008]). It therefore served as the reference for a nature-near, less disturbed forest ecosystem. This area includes a private reserve with low human impact resulting primarily from ecotourism in the form of birdwatchers. Most of the area within this sub-watershed is protected under the REGUA Association, which is financially supported by the Brazilian Atlantic Forest Trust (BART) with the stated objective of protecting the Atlantic Rainforest of the upper Guapiaçu river watershed. This is mainly done by enlarging protected areas through land purchase.

- In contrast, the sub-watershed Batatal represents a mixed system of the most relevant land uses with a mosaic of forest fragments (69%), pastures (28%) and agriculture (4%) considering the land use classification by Fidalgo et al. ([2008]). Predominantly, banana (perennial) is found in higher elevation areas, followed by annual crops mainly found in flat areas or lowlands (cassava, maize, beans and vegetables). Forest fragments in different stages of ecological succession are found in high elevation areas.

- In the Caboclo sub-watershed, the predominant land use type is forest (81%), followed by pasture (14%) and agriculture (3%) according to Naegeli ([2010]) and Fidalgo et al. ([2008]). Agricultural systems with a considerably higher intensity than in Batatal are found in this sub-watershed, mainly along the floodplain. The most common annual crops are maize, cassava, beans and vegetables, whereas perennial crops are rare. The higher elevation area is within the boundaries of the State Park Três Picos, where agricultural activities are carried out within the buffer area of this Park.

Figure 3: Sub-watershed landscapes of three rivers: Manuel Alexandre (left), Batatal (upper right) and Caboclo (lower right).

Both sub-watersheds, Batatal and Caboclo have undergone historical exploitation cycles, resulting in high deforestation, forest fragmentation and degradation, as well as intense soil erosion (Nehren et al. [2013]).

Methods to Assess Supply of Watershed Services

To calculate the opportunity costs (OCs) related to the provision of watershed services under varying land use systems, we carried out a cost-benefit analysis of representative farming systems in the region. For this, we developed detailed individual budgets for all activities within a given farming system (for definition see Beets [1990]). Activity budgets summarised revenue and cost information and were finally aggregated to calculate the average rate of return for each land use type (WBI[2011]). Crop budgets were compared for coherence with official current production costs used by the Rural Extension and Technical Assistance Agency (EMATER) of the state of Rio de Janeiro.

Our target population was households practicing some degree of agriculture at the sub-watershed level considering the sub-watersheds of Batatal and Caboclo. A household survey was carried out in two field campaigns in 2011 and 2012 with the permanent support of key local producers, EMATER of the municipality of Cachoeiras de Macacu, the City Council of this municipality and Embrapa Soils scientists. Expert interviews were carried out with the Director of the REGUA Reserve and other staff members in Manuel Alexandre sub-watershed, to better understand land use history and recent management practices in the region.

To define the sample universe, we created an inventory using the indirect census technique following Forero ([2002]). This process consisted of a participatory mapping exercise based on recent aerial imagery provided by the City Council of Cachoeiras de Macacu. This enabled us to assemble a list of all farm units within the sub-watersheds, which was confirmed by extensive field visits and supported thoroughly by local experts including a member of the Agricultural Department of the City Council (Cachoeiras de Macacu), the President of the Faraó Farmer's Association (A.L.A.F.) in the sub-watershed of Batatal, a member of the Rural Workers Union (in the sub-watershed of Caboclo) and the Director of the REGUA Reserve (Manuel Alexandre sub-watershed). As a result, a total of 32 households in Batatal and 60 in Caboclo were identified, of which 78 households within the two populated sub-watersheds were interviewed using a semi-structured survey. No interviews were made in the reference site of Manuel Alexandre sub-watershed. The sample size obtained is supported by Angelsen et al. ([2011]), who suggest a minimum sample size of 25–30 households from each community. This is valid for communities with 100 to 500 families. The designed questionnaire was based on various scientific publications and reports (see Rodriguez Osuna [2013]; Angelsen et al. [2011]; WBI [2011]; Gaese [2009]; Instituto Terra Mater [2009]; Forero [2002]).

Throughout the course of our fieldwork, two survey rounds were carried out. The first survey round included a random sampling of farm units within each sub-watershed to define "representative farming systems". Important selection criteria for these farming systems as suggested by Zimmer et al. ([2009]) and local experts included mainly: farm size, land tenure, production programme and agricultural management practices, and average location of farms in terms of metres above sea level (m.a.s.l.).

Once such farming systems were defined, a second survey round was launched to explore in-depth characteristics of farming systems with special attention to the inputs and outputs that are relevant to profitability among such systems.

In the sub-watershed of Batatal, we divided the farming systems by location in upland versus lowland, since this division significantly affects production patterns. Farm units in the uplands are located at an average altitude of 344 m.a.s.l. in contrast to those in the lowlands located at ca. 83 m.a.s.l. Agricultural production in the lowlands of Batatal is comparable to those sub-watersheds located along the Macacu River. The same occurs in Caboclo, which is representative for sub-watersheds along the Guapiaçu River (Figure 2).

Subsequently, a cost-benefit analysis was carried out for each farming system. The occurrence of each farming system was estimated and validated through local expert consultation as suggested by Angelsen et al. ([2011]).

Our sampling strategy focused on capturing the diversity of smallholder production systems in the region, yet our total sample size was too small to obtain a representative sub-sample of the large cattle operations that dominate in the lower part of the watershed. For the cost-benefit analysis of cattle production systems, we thus relied on additional in-depth interviews with a group of livestock producers deemed representative by officers of EMATER.

Based on interviews with selected livestock producers and secondary data on livestock systems in this area (see Quintana [2012]), we calculated livestock activity budgets for three slope categories: 1) ≤15°; 2) 16-25° and 3) >25°. These budgets were calculated under the assumption that profits for livestock production decreased with increasing slope, because of lower productivity of pasture, among other factors. This assumption is based on interviews with local farmers.

Once profits for agricultural and livestock systems were obtained, they were extrapolated to the watershed level using a Landsat based land cover classification that identified "agriculture" and "pasture" areas (Pedreira et al. [2009]; Fidalgo et al. [2008]). The agricultural profit calculated in the selected sub-watersheds (only considering lowland areas) was applied to all sub-watersheds in the same river network. In the sub-watershed of Batatal, we divided farming systems located in uplands and lowlands since this division influences significantly production patterns. This was not necessary in the Caboclo sub-watershed, where all farming systems are located in the lowlands.

The profit derived from farming systems is equal to the opportunity cost of converting agricultural or pastoral lands to forest, thereby reducing turbidity. For example low OCs are associated with low profits from current land use. Per hectare OC estimates for each sub-watershed thus represent the weighted average per hectare profits from the respective land cover types.

We relied on a spatial analysis of the vulnerability of water resources in the study area (Ferreira[2012]), which was understood as the likelihood of watershed service loss. In this case, our assessment considered state and pressure indicators following the Driving Forces, Pressure, State, Impact and Response (DPSIR) framework (see Borja et al. [2006]), where 50% were state indicators including: geomorphology, hydrogeology, drainage density, soils, index of circularity, index of areas of permanent protection (APP) fragments and slope. APP areas are established by the Brazilian Forest Code (Federal Law 4771/1965) and are defined as "protected areas, both covered or not with native vegetation, that have the environmental functions of preserving water resources, landscapes, geological stability, biodiversity, and genetic fluxes of flora and fauna, as well as protection of the soil and securing the wellbeing of human populations" (Ministry of Environment [2005]). These areas include a minimum vegetation area to protect riverbanks and headwaters. The other 50% included pressure or anthropogenic indicators such as phosphorous (P) and nitrogen (N) production, land use and road density (Ferreira [2012]). All mentioned factors (state and pressure) were weighed by hydrological expert consultation by Ferreira [2012] (Figure 4). Single indicators were based mostly on published official maps predominantly generated by Embrapa Soils.

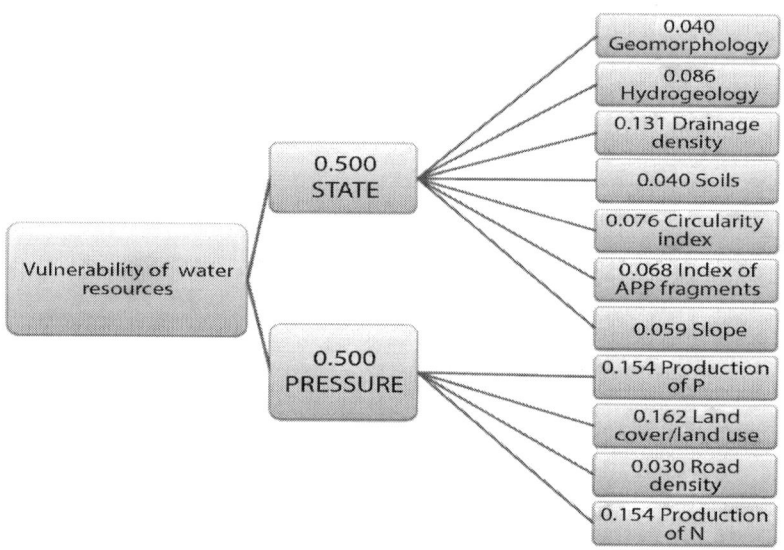

Figure 4: State and Pressure Indicators to assess water resources vulnerability. These indicators were weighed by hydrological expert consultation to assess the vulnerability of water resources in the Guapi-Macacu watershed (Modified from Ferreira [2012]).

Our next step was to identify priority areas for watershed service provision. For this, we overlaid environmental and economic criteria, i.e. vulnerability of water resources and spatial OCs within the Guapi-Macacu watershed. This allowed us to identify the areas where low OCs of shifting land use towards improving watershed services can result in high watershed service payoffs. These sub-watersheds were given the highest priority for intervention with watershed service improvements. Second and third priorities were given to sub-watersheds with high OCs and vulnerability and those with low OCs and vulnerability, respectively. The latter is based on the assumption that environmental goals are more important than cost criteria. However, this could be changed when there are budget restrictions and when there is intent to increase efficiency of payments in compensation schemes for watershed services. The lowest priority areas are those sub-watersheds with high OCs and low vulnerability of water resources.

Methods to Assess the Potential Demand for Watershed Services

Demand for watershed services was assessed by identifying costs related to the end-user of those services (Honey-Roses et al. [2013]). For this particular study case, it included quantifying water treatment costs incurred by the main watershed user, the state water utility (CEDAE). Water treatment costs that may be avoided if forests are restored can be translated as the potential willingness to pay (WTP) for watershed services. For example, if a one unit reduction in turbidity levels implies 10 additional monetary units in treatment costs, the water company's maximum WTP for watershed services will be 10.

We applied the avoided cost method focusing on the annual operational costs of chemical products for the treatment of raw water in the period between 1998 and 2011 from the local state water utility company. This approach required identifying key water quality indicators related to the main operational cost categories of the water utility company.

Based on expert interviews and as suggested by Medeiros et al. ([2011]), Reis ([2004]) and Dearmont et al. ([1998]), we identified turbidity as the key indicator amongst all water quality parameters, since an increase in turbidity implies an increased concentration of suspended solids in surface waters and is likely to reduce the quality of raw water to be treated by the water utility. The conversion of forest to other land uses such as agriculture or pasture caused by farming systems can result in higher turbidity values, which in turn correlate with higher water treatment costs to reduce the concentration of suspended solids for public water supply (Medeiros et al. [2011]; Reis [2004]; Dearmont et al. [1998]).

In addition Medeiros et al. ([2011]) and Reis ([2004]) suggest that chemical products account for close to 60% of annual operational costs in treatment plants in São Paulo (Reis [2004]). These products are used to flocculate the suspended particles, measured as turbidity, that are found in raw water to fulfill regulations on drinking water quality for human consumption (maximum value of 5 NTU).

RESULTS

Vulnerability of Water Resources

Water resource vulnerability is a function of both anthropogenic impact/pressure indicators and of environmental state indicators, so the assessment of vulnerability must account for this set of indicators. When only considering environmental state indicators, sub-watersheds located in higher areas of the Guapi-Macacu watershed tend to have higher vulnerability to anthropogenic pressure than those in the lowlands (Ferreira [2012]). Pressure indicators in a watershed are highly influenced by population density, land use practices, presence of urban settlements, road density and other factors previously mentioned. Therefore, the sub-watersheds with relatively high anthropogenic impacts are those with a high density of urban settlements and rural population nuclei. Sub-watersheds found in higher areas of the watershed have considerably lower impact values (Ferreira[2012]).

Figure 5 shows the sub-watersheds with higher vulnerability in the darker tones and those with lower vulnerability in the lighter tones. The less vulnerable sub-watersheds are found in the lower areas of the watershed and one of these is considered a natural protected area with limitations and restrictions in land use, despite its lower vulnerability.

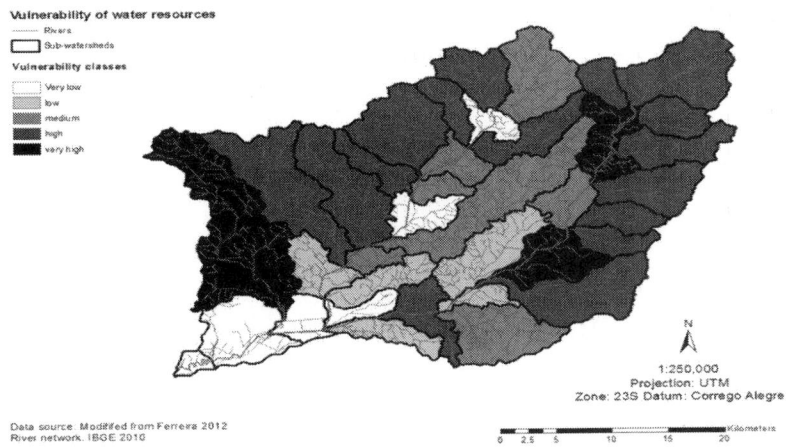

Figure 5: Vulnerability of water resources in the Guapi-Macacu watershed (Ferreira [2012]).

Agricultural Production and Opportunity Costs

Field surveys carried out in the Batatal and Caboclo sub-watersheds showed differences in production patterns and differences in specific environmental factors that reflect both the effect of farming systems on the provision of watershed services as well as distinct farming systems' profitability.

Relevant characteristics that differ considerably across these two sub-watersheds include land tenure and land use. In Batatal, most of the land is obtained by heritage or with a provisionary land title (66%), followed by banana share-croppers (19%), who are generally entitled to half the harvest. In contrast, in Caboclo most of the population lives within a settlement that keeps a common forest area called a "private reserve". This reserve is protected and is restricted from cultivation. Therefore, 85% of the farm-households in Caboclo are located in the lowlands, while the rest are found on the hillsides. In both sub-watersheds, one farm unit does not surpass 14 hectares, which is considered typical for family-oriented agriculture for the municipality of Cachoeiras de Macacu (MDA [2010]). After considering the specific differences among farming systems in Batatal and Caboclo, four types of farming systems (FS) were classified in Batatal and two in the sub-watershed of Caboclo (Table 3).

Table 3: Summary of annual farming systems profits in the study area

| Main crops | Unit | Batatal sub-watershed | | | | Caboclo sub-watershed | | | Pasture land systems oriented to beef production considering slopea | | |
| | | Upland | | Lowland | | Lowland | | | | | |
		FS1	FS2	FS3	FS4	FS5	FS6a	FS6b	≤15°	16° to 25°	>25°
Banana	(BRL)	972	972	983							
Cassava			-1,334	3,950	3,950	5,231	5,232	5,232			
Green maize (summer)				2,371	2,371	2,066	2,067	2,067			
Green maize (winter)				3,438	3,438	2,371	2,371	2,371			
Yam				2,861	2,861	6,199	6,199	6,199			
Courgette					3,633						
Gilo								10,941			
Okra							10,092				
Beans (summer)						1,419	1,419	1,419			
Beans (winter)						1,522	1,523	1,523			
Occurrence	(%)	80	20	50	50	70	15	15			
Typical area and crop distribution for each farming system (FS)	(ha)	10.3	10.3 (8.3 banana, 2.0 cassava)	6.0 (2.0 cassava, 2.0 green maize, 1.0 yam, 1.0 banana)	6.0 (2.0 cassava, 2.0 green maize, 1.0 yam, 1.0 courgette)	4.0 (1.0 cassava, 1.5 green maize, 1.0 yam, 0.5 beans)	4.0 (1.0 cassava, 1.5 green maize, 0.5 yam, 0.5 okra or gilo, 0.5 beans)				

Per hectare profit	(BRL ha−1)	972	526	3,894	4,336	4,889.7	5,376.3	5,482.5	20	40	100
Average value for extrapolation at the watershed level	(BRL ha−1)	Macacu River lowland				Guapiaçú River					
		4,114.8b				5,051.6			20	40	100

aAccording to expert interviews, there is a gain of 40 to 60 kg of live animal weight in high slope areas in contrast to those in lowlands with a gain in live animal weight of 120 to 150 kg in this particular watershed; bSince upland agriculture in the Batatal sub-watershed (FS1 and FS2) was not detected by the available land use classification (Naegeli[2010]; Pedreira et al.[2009]; Fidalgo et al.[2008]), we were limited to using average per hectare annual profits for lowland agriculture in Batatal (FS3 and FS4) resulting in an average annual per hectare value of 4,114.8 BRL.

Rodríguez Osuna et al.

Rodríguez Osuna et al. Ecological Processes 2014 3:16   doi:10.1186/s13717-014-0016-7

In Batatal, we found that 80% of FS in the uplands are specialised in banana (Musa sp.) production (classified as FS 1) and 20% had a mixed system of cassava (Manihot sp.) and banana (FS2). In the lowlands of Batatal, two additional FS where agriculture is more intensive were classified as FS3 and FS4 and are equally distributed. FS3 has a production system composed of cassava, green maize (Zea mays), yams (Colocasia sp.) and courgette (Cucurbita sp.), while FS4 has the same mix of cassava, green maize and yams, but banana instead of courgette (Table 3).

Cassava is the dominant crop in the Caboclo sub-watershed, followed by green maize, yams and common beans (Phaseolus vulgaris). Cassava is the most cultivated crop, mainly due to low investment requirements and because of relatively stable returns after a cropping period of 8 to 9 months. Green maize requires higher investments; however, it provides relatively rapid returns after only 90 days. Common beans are used to improve soil fertility (nitrogen fixation) and as an alternative to the other products. Other relevant short-cycle products are okra (Hibiscus esculentus) and gilo (Solanum gilo).

The most common farming system in the Caboclo sub-watershed (FS5) combines cassava, yams, common beans and green maize (70%), while the rest (FS6-FS6a) combine cassava, yams, common beans, green maize and -additionally- okra or gilo (Table 3). Typically, green maize, courgette and beans are planted twice in one cropping cycle.

Agriculture in the uplands of Batatal is clearly less intensive in fertilizer use than in the lowlands, especially given that banana production in the uplands is carried out without fertilization. The remoteness of these locations makes intensive production less attractive than in the lowlands. In the lowlands of Batatal, vegetable producers applied on average of 240 kg of fertilizer per hectare each year. In Caboclo, more than 70% of households used fertilizers for their agricultural production, at an average of 547 kg per hectare each year.

In the uplands of Batatal, all households used herbicides as their basic approach to weed control, whereas a variety of agrochemicals were used in the lowlands. The high use of herbicides for weed control could be the result of the shortage of and high cost of labour that would otherwise undertake this activity. It should also be noted that the Atlantic Forest Law (Law 11.428/1986) and the Brazilian Forest Code (Law

12.651/2012) bring certain limitations to agricultural production. The Atlantic Forest Law bans the conversion of secondary forest into land uses such as agricultural land. As an example, a collapse of the market price for banana around fifteen years ago left many banana plantations uncultivated and secondary forests developed and expanded, leading to abandonment of these plantations. In addition, the Brazilian Forest Code defines for the Atlantic Forest Biome that 20% of rural properties need to be maintained as a permanent forest reserve "Reserva Legal". The Brazilian Forest Code also prohibits the clearing of primary vegetation on steep slopes (>45°), along the margins of rivers and streams and in headwater (source) areas, which are classified as areas of permanent protection (APPs) (Ministry of Environment [2005]).

In Batatal, the per hectare average annual profit was estimated at 4,115 BRL and in Caboclo at 5,052 BRL (Table 3). Returns for agriculture tend to be higher in the lowlands, where the intensity of production is higher than in the uplands, and there is a higher use of agricultural inputs especially fertilizers for cash crops.

According to local expert interviews small livestock farmers are considered those with farm size less than 20 hectares, while big-scale producers are considered those with more than 400 hectares. According to official cattle vaccination data in the Municipality of Cachoeiras de Macacu in 2011, an estimation of the herd size can be given on the base of vaccinated animals. This resulted in 27,995 animals in all three districts of this Municipality (Secretary of Agriculture, Cattle Farming, Fisheries and Supply [2011]). Most livestock farmers (90%) have less than 500 animals and small-scale producers are considered in this municipality those with less than 20 animals. Mostly, animals are distributed in paddocks without dividing fences.

Livestock production systems achieved profits of 20, 40 and 100 BRL per hectare annually, depending on slope class (Table 3).

Area-weighted OCs per sub-watershed were spatially mapped, ranging from 14 to 1,660 BRL per hectare (Figure 6). This reflects that extensive pasture is the most important land use in the Guapi-Macacu watershed according to area, which is an activity with comparatively low per hectare profits.

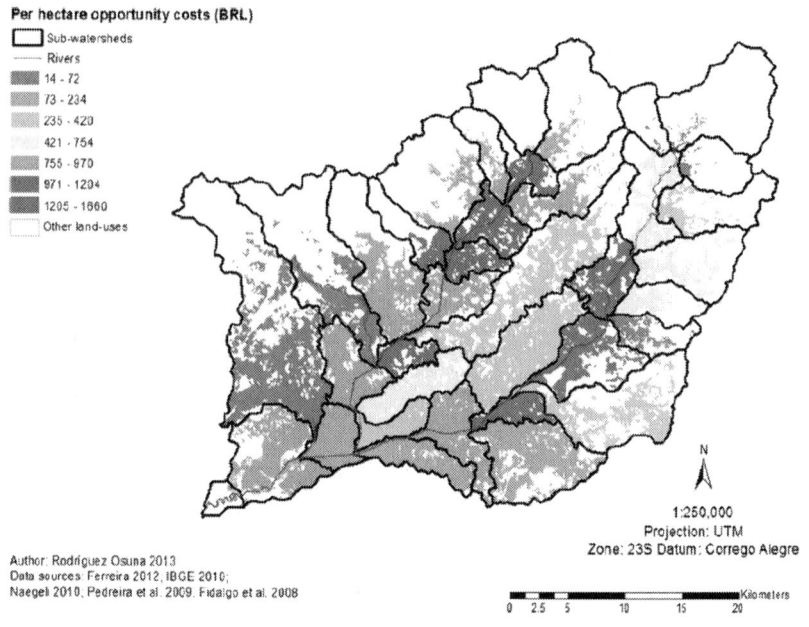

Figure 6: Spatial per hectare opportunity costs in BRL for the land uses "agriculture" and "pasture" within each sub-watershed in the Guapi-Macacu watershed. The white areas within the map correspond to those with land uses other than agriculture in the Guapi-Macacu watershed.

Figure 5 shows that quite a number of high OC areas are located close to the main two river beds, where intensive agriculture predominates. Resulting OCs from agriculture occur only in lowlands and close to the river plain, which have higher nutrient concentration than in higher slope areas.

Low OCs areas are often either located in the steep slope areas of the upper watershed or in the valleys, where extensive pasture areas dominate. However, these areas are also found in the lower parts of the watershed, where a small-scale settlement promoting family-oriented cattle ranching was launched a decade ago by the government of Rio de Janeiro.

Forests predominate mainly in higher slope areas (mostly in white in Figure 6), which originate, to a great extent, in protected areas such as the State Park "Três Picos", the National Park "Serra dos Órgãos" and the State Ecological Station "Paraíso".

Analysis of Environmental and Economic Criteria for Watershed Service Conservation and Improvement

Analysis of environmental (vulnerability of water resources) and economic criteria (OCs of watershed service provision) in the Guapi-Macacu watershed allowed us to identify priority areas for watershed service conservation and improvement. These areas are the land use-based management options (i.e. conversion of pastoral or agricultural lands to forest) with the highest potential of improving water quality and lowest OCs (Figure 7).

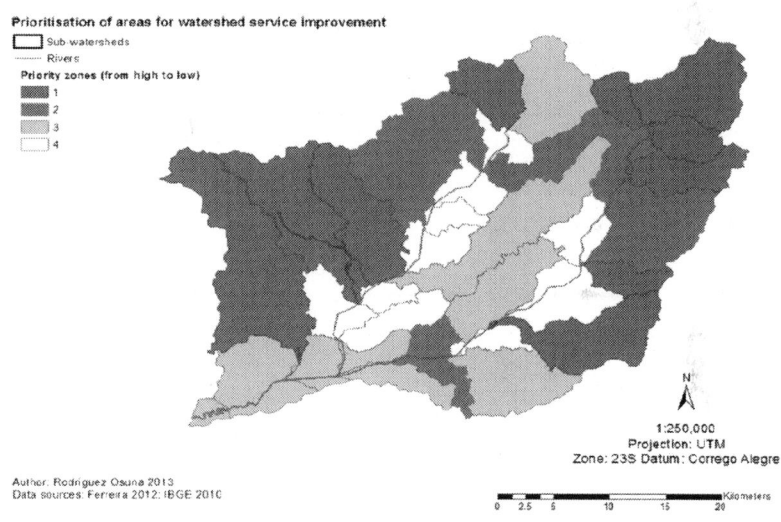

Figure 7: Priority areas for improving or ensuring watershed services of the Guapi-Macacu watershed. Priority areas for improving or ensuring watershed services (water quality) where 1 is considered high and 4 represents low priority.

Our results highlight the sub-watersheds where the vulnerability of water resources are highest and OCs of converting land uses to foster watershed service provision are lowest. The high priority areas (in dark red) are in most cases found in sub-watersheds with steeper average slope levels where impact on water resources was found to be high (Figure 7).

Demand for Watershed Services: Water Treatment

According to micro-economic theory, we interpret the water utility company's demand for chemicals to treat water as its willingness to pay for a desired water quality level (in this case turbidity under 5 NTU for human consumption). The treatment cost of an additional turbidity unit is thus equivalent to the company's potential willingness to pay for any measure that reduces turbidity by the same amount (see avoided cost method, for example, in Perman et al. [2003]).

Based on expert interviews and relevant studies (Medeiros et al. [2011], Reis [2004] and Dearmont et al. [1998]), which identified turbidity as the key water quality parameter relevant for water treatment costs, we applied the avoided cost method to the local water utility company. In Table 4, the main characteristics of the water treatment are presented.

Table 4: Characteristics of water treatment utility in the lower catchment part of the Guapi-Macacu watershed in 2011

Treatment unit: Laranjal - CEDAE	Municipality: São Gonçalo
Captivation area from the GMW	1,263 km2
Forest cover in the watershed	48.8%
Treatment type	Conventional
General treatment phases	Captivation, sedimentation, coagulation, flocculation, decantation, filtration, disinfection, water fluoridation and pH correction
Treated water flow	Average flow in 2011, 5.35 m3s−1
Population supplied with treated water	2,000,000 inhabitants
General chemical products used	Aluminium sulphate $Al_2(SO_4)_3$, polyelectrolyte, hexafluorosilicic acid H_2SIF_6, chlorine, calcium oxide CaO

Costs with chemical products for the treatment of raw water	2.31 × 106BRL (year 2011)
Turbidity of raw water, average values	17.10 NTU
Treated water characteristics	Colour, 2.50 uH
	Turbidity, 0.34 NTU
Total costs with chemical products and electricity in a year for the water treatment unit	For the treatment of 6 m3s−1, • 300,000 [BRL] (monthly expenditure for chemical products)
	• 100,000 [BRL] (monthly expenditure for electricity in the captivation and production area within the unit)

Rodríguez Osuna et al.

Rodríguez Osuna et al.Ecological Processes 2014 3:16 doi:10.1186/s13717-014-0016-7

Based on data from CEDAE (treatment unit Laranjal) for the period between 1998 and 2011, we estimated the avoided costs of a reduction of 1% turbidity at 15,510 BRL, considering an average volume of treated water of 174,545 cubic metres at an average cost of 22.2 BRL per 1,000 cubic metres (Table 5). The Pearson correlation coefficient for average annual turbidity levels and costs for chemical products in this water utility for the period between 1998 and 2011 was 0.4. This correlation value is relatively low compared to other studies carried out in Brazil. For example, Reis ([2004]) calculated a correlation value of 0.7 for seven water treatment units in São Paulo and Cabral de Sousa ([2011]) found a coefficient of 0.9 when comparing 10 different sub-watersheds and their treatment costs in São Paulo.

Table 5: Annual avoided costs from 1998–2011 for the local water utility company

Average quantity (m3yr−1)	Average cost (1,000 BRL m−3)	Total cost (BRL)	Avoided cost for 1% of turbidity reduction (BRL)
174,545.3	22.2	3,877,482.0	15,509.9

Rodríguez Osuna et al.

Rodríguez Osuna et al. Ecological Processes 2014 3:16   doi:10.1186/s13717-014-0016-7

Reis ([2004]) found that chemical treatment costs in water utility units on the Piracicaba River are 12.7 times higher than the cost of treating water from the Cantareira system. This author argues that this finding can be explained by the considerably lower forest cover in the Piracicaba watershed (4.3%) compared to the Cantareira watershed (27.2%).

Therefore, the geographical location of forests, as well as land use, soil type, geomorphology and predominant geology are considered relevant factors that influence water quality from headwaters (springs) and water treatment costs for public supply. Nevertheless, Reis ([2004]) showed that the percentage of forest cover is often a sufficiently informative indicator of watershed health and thus water quality.

Water quality monitoring obtained in seven monitoring campaigns in the years 2010 and 2011 along the whole watershed demonstrated lower turbidity levels in the highest parts of the sub-watershed, where forest cover is higher. For example, our reference sub-watershed Manuel Alexandre presented an average value of 0.8 NTU at the outlet as compared to 17.4 NTU at a lower region close to the water intake point of the water utility (Paiva et al. [2011]).

Is paying for land use changes that foster watershed services cheaper than treating water?

The estimated OCs of converting land use/land cover to the benefit of water quality can range between 4,000 to 5,000 BRL per hectare each year for agricultural systems and less than 100 BRL for pasture land. However, the actual OCs per hectare in many sub-watersheds are likely to be much lower, especially if land close to rivers and headwaters is covered by extensive pastures. On the demand side, we estimate that the water company CEDAE's maximum WTP for land use change in the watershed based on avoided costs is 15,510 BRL per additional 1% reduction in turbidity levels at the water intake point of the water utility.

Ideally, a full-scale hydrological model for the watershed would provide us with the potential effect of alternative land use scenarios on turbidity at the water intake point of the water utility. In the absence of such a model, we can only provide an informed estimation with regard to the viability of PWS in the watershed. At the high OCs end, it is clearly unrealistic to expect that the conversion of less than 3 hectares of intensively used cropland in the whole watershed (the avoided cost of water quality reduction: 15,510 BRL / maximum per hectare OC: 5,482.5 BRL per ha = 2.83 ha) will result in a 1% reduction of turbidity levels. In contrast, carefully selected pasture and low intensity agricultural sites could potentially be converted into forest (162 to 814 hectares) in the case of pastures. Land use changes at that scale are more likely to bring about measurable changes in turbidity levels if located in zones with a large impact on river water quality.

Unfortunately, given the spatial distribution of pastures and high intensity agriculture in the watershed, there are likely to be limited opportunities to convert large tracts of land at low costs. Payments for forest recuperation may thus likely remain a complementary watershed management measure in our study area. As an alternative to full scale conversion, some simple pasture management techniques, such as limiting access of cattle to the riverbed in lowland pastures (see dark red areas in the centre of Figure 7), could prove to be comparatively low cost and highly effective measures to reduce turbidity levels close to the water intake point of CEDAE.

Inadequate livestock grazing practices can compromise water quality to the point where is considered degraded and highly polluted and not able to meet water quality standards (EPA [2013]). Therefore, excluding livestock from streams and improving range management practices can contribute to reduce turbidity on streams (EPA [2013]).

Although the water supply company's WTP does not match the estimated OCs, payments from other water users are an additional option that can be taken into account in the design of a PWS scheme in this watershed. Particularly important would be the demand of water by COMPERJ (the Rio de Janeiro petrochemical complex). In addition, this assessment was solely carried out on the basis of water quality improvements in terms of turbidity levels, since it was found this service to be relevant for the demanders downstream. However, accounting for additional ecosystem services provided by forest ecosystems would increase the potential WTP for ecosystem services in this region.

DISCUSSION

This study estimated the costs involved in both supply and demand of water quality maintenance and improvement; where we identified priority areas for supply in order to target watershed management measures or support the launch of compensation schemes such as PES or PWS. We estimated demand by assessing the willingness to pay for a water quality improvement in terms of turbidity by the main watershed user, the state water utility company (CEDAE). This approach showed to be helpful for the design of watershed payment schemes in other local contexts (Pagiola et al. [2010], Martinez de Anguita et al. [2011], Garcia-Nieto et al. [2013], Martin-Ortega et al. [2012]). Adjusting off-the-shelf modelling packages, such as SWAT, for our purposes (see for example, Martinez de Anguita et al. [2011] and Quintero et al. [2009]), was deemed inappropriate by local hydrologists. Attributing water quality dynamics to land use changes is one of the most challenging issues in ecosystem services research (De Groot et al. [2010]). Since we lack an appropriate hydrological model, we have left the quantification of the effects of specific land use changes on water quality improvement in our study area for future research.

Some limitations of this study are related to the use of turbidity as a water quality indicator. Land use is most likely the most relevant factor that influence turbidity changes in streams (ECI [2014]). However, turbidity is subject to natural physical and biological variations even if the watershed is forested. The physical factors relate particularly to erosion processes, landslides, and mudslides after heavy rainfalls. These accelerated slope processes that are associated with high sediment loads and turbidity can even take place under closed forest cover, as seen for example during the 2011 mudslides and floods (Nehren et al. [2014]). The nature of soils and geology in the watershed (e.g. erosion of the riverbed) also determines how easily erosion might occur (ECI [2014]).

We found a lower correlation between water treatment costs and turbidity (0.4) than the 0.7 found by Reis ([2004]) in her analysis of 7 treatment utility plants in São Paulo. The use of average values for turbidity and the costs of chemicals used for treatment would most likely mask a higher correlation. However, our correlation is relevant since it supports our conclusion that the WTP from the water supply company is likely to be rather low. Yet, turbidity levels monitored at the water intake point of the water supply facility do not exceed 50 NTU (Paiva et al. [2011]).

Comparing these turbidity levels for water treatment with other cases in Brazil, we found mean turbidity levels of 16 NTU and 64 NTU in São Paulo (the first value between April to October and November to March respectively) at the water intake point of the River Piracicaba. In this particular case, it was found that this treatment plant had the highest costs related to the use of chemical products for water treatment, which in addition might have been correlated with the low forest cover of 4.3% (Reis [2004]). During these same periods mean turbidity in the Cantareira system (dam) varied from 3–9 NTU respectively with the lowest costs for chemical products with a more elevated forest cover 27.1% (Reis [2004]).

However, with regards to the use of chemical products for water treatment and their relative costs, it should be noted that not all chemical products are used to tackle turbidity. The main product used to flocculate suspended material and reduce turbidity is aluminum sulphate, which has the highest share of the cost of chemicals for the treatment process. The use of the other single chemical products is also correlated to the performance of the treatment process, where, for example the better the quality of the final product (i.e. treated/filtered water), the lower the cost with the rest of the chemical products, namely hexafluorosilicic acid, chlorine and calcium oxide.

Agricultural land use clearly represents the most important entry point for watershed managers in the studied area, which is why we focused on quantifying the OCs of changing rural land cover. It should be noted that Lorz et al. ([2011]) demonstrated that urban settlements also negatively impact water quality in terms of turbidity, ammonium (NH_4^+) and Chemical Oxygen Demand (COD). Therefore, future studies should also consider the costs of changes to waste water treatment in upstream urban settlements.

Moreover, we note that the land cover classification used in our analysis does not distinguish between agricultural production systems and underestimates the total land under production due to low resolution (Pedreira et al. [2009]; Fidalgo et al. [2008]). Land use classifications with higher resolution (based on SPOT 5) only exist for the upper part of the Guapi-Macacu watershed, but such data would permit our OC analyses to be extrapolated in more detail.

Comparing our OC estimates to actual transfers in existing PWS schemes in the Atlantic Forest, we find that per hectare payment values range from 10 BRL annually to 566 BRL monthly (Veiga and Galvadao [2011]). These include annual per hectare payments of 176 BRL in the municipality of Extrema, state of Minas Gerais; 25–125 BRL in the watershed of PCJ in the state of São Paulo; 10–60 BRL in the watershed of Guandu, state of Rio de Janeiro; 80–340 BRL in the watershed Benevente, state of Espírito Santo; 80–340 BRL in the Guandu watershed, states of Espírito Santo and Rio de Janeiro; 75–563 BRL paid by the OASIS Foundation, states of Espírito Santo and Paraná; and 175–577 BRL monthly payments in the state of Santa Catarina (Veiga and Galvadao [2011]).

At the regional level in Latin America, we can compare our estimates to well-known schemes, such as the Costa Rican PES programme (FAO [2007]). Here, annual per hectare payments vary from 98 BRL for natural regeneration to 152 BRL for forest preservation and more than 2,332 BRL for new forest plantations for a time period over five years (Ecosystem Marketplace [2010]). The National Programme for Hydrological Environmental Services in Mexico (PSAH) pays between 57–90 BRL per hectare annually depending on the forest type considering the OCs of land conversion (Ecosystem Marketplace [2010]; Muñoz-Piña et al. [2008]). The Ecuadorian PWS Programme in Pinampiro pays landowners around 14–29 BRL per hectare annually (Wunder and Alban [2008]), while the Los Negros programme in Bolivia pays in-kind with beehive boxes for honey production (Asquith and Wunder [2008]). Agricultural OCs in our study area thus clearly exceed the average annual payment in existing PWS schemes by more than a factor of 10.

However, we note that payments made under the above mentioned schemes may focus in other actions other than conversion from agriculture or pastoral lands into forest. In the case of the Atlantic Forest schemes, mostly payments are directed to protect headwater areas and restore and conserve riparian forests. However, actions other than conversion from agriculture or pastoral lands into forest can be paid for (eg. soil conservation practices). The price paid to farmers in the PSAH Mexican scheme has been decided based on the assumption that corn production would be an alternative to conservation (Ecosystem Marketplace [2010]). The Pinampiro PWS programme pays for forest and páramo protection and regeneration (Wunder and Alban

[2008]), while the Los Negros programme compensates for avoided deforestation of cloud forests (IIED [2012b]).

Our results may prove useful for ongoing watershed conservation initiatives, such as the World Bank funded Rio Rural carried out by the State Secretary for Agriculture and Livestock in the state of Rio de Janeiro (RioRural [2013]). This and many other initiatives in the Atlantic Forest region have shown great interest in incentive-based watershed management approaches. Furthermore, Brazilian legislation has been supporting such approaches, such as the Brazilian National Law on Water Resources (Law 9433/1997), which allows for example the establishment of watershed service markets and permits charging for water use as it is taking place in some watersheds in Rio de Janeiro (Veiga [2008]).

Equally relevant is the more recent Law 9985/2000 "SNUC" that adopts the protector-receiver principle, which allows for rewards in exchange for good natural resource stewardship (Strobel et al. [2007]). Funding for incentive-based watershed management may be derived from Brazil's "ecological" tax system, which compensates municipalities for conservation costs based on the value-added tax (ICMS) (Marques [2009]; Veiga [2008]).

In addition, the decree No. 42029/2011 established the Payment for Environmental Services mechanism under the Rio de Janeiro State Programme for Conservation and Revitalization of Water Resources (PROHIDRO). This decree has given legal status to the PES Programmes already underway in the state of Rio de Janeiro, such as the Lagos São João Committee through Good Practices Fund (Funboas) and Guandu Committee through Water and Forest Programme implemented in the municipality of Rio Claro.

This study can potentially contribute to implementation of these laws in the Guapi-Macacu watershed in a more effective way with respect to social, economic and environmental aspects. Given that our study results take into account these different aspects and are based on primary data obtained in the watershed, our study can contribute by identifying limitations and opportunities related to economic watershed compensation programmes.

CONCLUSIONS

Assessing the economic scope of incentive-based watershed management requires knowledge of both the costs of providing additional watershed services through land use and land cover change (service supply) and the willingness to pay (demand) for such services. This study has quantified these two necessary inputs to enable an informed decision-making process in the context of the Guapi-Macacu watershed in the state of Rio de Janeiro, Brazil.

With regard to supply, we found that land users' OCs range between 972–5,482 BRL per hectare for cropland and <100 BRL per hectare for pastures. At the sub-watershed level, zones that use a high proportion of pasture but low proportion of cropland in the land use mix will accordingly have low area-weighted average OCs and vice versa. With regard to demand, we assessed water treatment costs and found a 1% reduction in water turbidity levels to be worth 15,510 BRL on average to the water company, CEDAE.

In absolute terms, only a relatively small share of land in the watershed is cropland (5,560 hectares); the area covered by pasture (52,374 ha) and forest (61,665 ha) is much larger. However, cropland covers a considerable amount of land with potential to provide watershed services and, as a result, high OCs become a critical limitation on strategies that aim to change land use for the benefit of water quality.

Nevertheless, given the spatial distribution of pastures and high intensity agriculture in the watershed, we find that payments for forest recuperation are likely to be cost-effective in sub-watersheds with a high proportion of pasture land. Some of these sub-watersheds also happen to exhibit high levels of water resources vulnerability and thus represent potential priority intervention zones for watershed management (Figure 6). In these areas, watershed planners may reap considerable benefits from active pasture management options, such as silvopastoral systems or strategic fencing of riparian areas. In addition, sustainable agricultural and soil conservation practices can bring additional benefits in terms of reducing and avoiding sedimentation, erosion, which in turn result in improving water quality.

To the extent that smallholders in the study area also depend on subsistence production for their livelihoods, land use based watershed management strategies will have to also consider food security issues, especially in the upper part of the watershed.

Our study area reflects watershed management issues that are typical across many parts of the Atlantic Forest region, which often have diverse agricultural mosaics and thus highly variable OCs. Wherever intensively used cropland dominates in the vulnerable zones of large watersheds, land use planners will find it difficult to rely on the use of payments for watershed services alone. Effective watershed management will then have to be combined with enhanced monitoring and enforcement activities to ensure that the Brazilian Forest Law is complied with, particularly with regard to riparian forests. Research designed with a bottom-up approach that quantifies the potential compliance costs of land users can clearly help policy makers to target both incentives and disincentives in a cost-effective manner.

AUTHORS' CONTRIBUTIONS

VRO undertook the field work, carried out the analysis and wrote the manuscript under her PhD thesis at the University of Leipzig. JB and RBP supported the conception of the research design and data analysis. JB, RBP and UN reviewed and commented on various versions on this manuscript. UN provided insights on ecosystem management aspects. HG and JH contributed to the overall research design and gave valuable comments on aspects related to farming systems analysis and landscape ecology (respectively). All authors read and approved the final manuscript.

ACKNOWLEDGEMENTS

The authors acknowledge financial support of the IPSWAT Programme (financed from the German Federal Ministry of Research-BMBF) and the DINARIO Project: "Climate Change, Landscape dynamics, Land use and Natural Resources in the Atlantic Forest of Rio de Janeiro". We thank the support of the Team of Researchers of MP2 Embrapa Soils of Rio de Janeiro and DINARIO. In our study region, we specially thank Lenilson Biazatti (from the Rural Workers Union), Nicholas Locke (REGUA Association Director), Thabta Matos de Mata from the City

Council of Cachoeiras de Macacu, Jocemir Da Silva (Chief technical extension officer of EMATER in Cachoeiras de Macacu), Demerval Pereira de Sousa (President of A.L.A.F.), technical staff from the state water utility company (CEDAE – Imunana Laranjal) and all farmers who enriched this study with their valuable time. We are also very grateful for the proofreading of this text by Dr. Kylie Quinn and Meghan Doiron.

REFERENCES

1. Angelsen A, Larsen HO, Lund J (2011) Measuring livelihoods and environmental dependence: Methods for research and fieldwork. Routledge, London.

2. Arriagada RA, Ferraro PJ, Sills EO, Pattanayak SK, Cordero-Sancho S (2012) Do payments for environmental services affect forest cover? A farm-level evaluation from Costa Rica. Land Econ 88(2):382-399

3. Asquith N, Wunder S (2008) Payments for Watershed Services. The Bellagio Conversations. Fundación Natura Bolivia, Bolivia.

4. Balvanera P, Uriarte M, Almeida-Lenero L, Altesor A, DeClerck F, Gardner T, Hall J, Lara A, Laterra P, Pena-Claros M, Silva Matos DM, Vogl AL, Romero-Duque LP, Arreola LF, Caro-Borrero AP, Gallego F, Jain M, Little C, de Oliveira XR, Paruelo JM, Peinado JE, Poorter L, Ascarrunz N, Correa F, Cunha-Santino MB, Hernandez-Sanchez PA, Vallejos M (2012) Ecosystem services research in Latin America: The state of the art. Ecosystem Services 2:56-70

5. (1998) Improving Water Utilization from a Catchment Perspective. SWIM Paper. International Water Management Institute, Colombo, Sri Lanka.

6. Beets WC (1990) Raising and Sustaining Productivity of Smallholder Farming Systems in the Tropics. A Handbook of Sustainable Agricultural Development. A Handbook of Sustainable Agricultural Development. AgBe Publishing, Holland.

7. Bennett G, Carroll N, Hamilton K (2013) Charting New Waters. State of Watershed Payments. Forest Trends, Washington. [http://www.springer.com/ environment/ environmental+management/ book/ 978-94-007-5175-0].

8. Boerner J, Vosti S (2012) Managing Tropical Forest Ecosystem Services: An Overview of Options. In: Muradian R, Rival L (eds) Governing the Provision of Ecosystem Service, vol 4. Springer, pp 21–46,

9. Borja A, Galparsoro I, Solaun O, Muxika I, Tello EM, Uriarte A, Valencia V (2006) The European Water Framework Directive and the DPSIR. A methodological approach to assess the risk of failing to achieve good ecological status. Estuar Coast Shelf Sci 66(1):84-96

10. Calder I (2005) Blue revolution: Integrated land and water resources management 2nd edition. Earthscan, London

11. (2001) Perfil do Ecossistema Mata Atlantica, Hotspot de Biodiversidade – Brazil. Critical Ecosystem Partnership Fund,   .

12. Costanza R, d'Arge R, de Groot R, Farber S, Grasso M, Hannon B, Limburg K, Naeem S, O'Neill RV, Paruelo J, Raskin RG, Sutton P, van den Belt M (1997) The value of the world's ecosystem services and natural capital. Nature 387(1):253-260.

13. Daily GC, Matson PA (2008) Ecosystem services: From theory to implementation. Proc Natl Acad Sci 105(28):9455-9456

14. de Cabral Sousa WJ (2011) Pagamento por Servicos Ecossistemicos: Mata Ciliar, Erosao, Turbidez e Qualidade de Agua: Produtos Tecnicos 1. Projeto de Recuperacao de Matas Ciliares, Brazil.

15. De Groot RS, Wilson MA, Boumans RM (2002) A typology for the classification, description and valuation of ecosystem functions, goods and services. Ecol Econ 41(3):393-408

16. De Groot RS, Alkemade R, Braat L, Hein L, Willemen L (2010) Challenges in integrating the concept of ecosystem services and values in landscape planning, management and decision making. Ecol Complex 7(3):260-272

17. Dean W (1997) With Broadax and Firebrand: The Destruction of the Brazilian Atlantic Coastal Forest. University of California Press, United States of America.

18. Dearmont D, McCarl BA, Tolman DA (1998) Costs of water treatment due to diminished water quality: a case study in Texas. Water Resour Res 34(4):849-853 [http://assets.panda.org/downloads/runningpurereport.pdf]

19. Dudley N, Stolton S (2003) Running Pure: The Importance of Forest Protected Areas to Drinking Water, The Arguments for Protection Series. World Bank/WWF Alliance for Forest Conservation and Sustainable Use, United Kingdom, . Accessed 10 Jan 2011.

20. (2014) Chapter 3 – Streams. In: Total Suspended Solids and Turbidity in Streams. Department of Ecology, State Washington. [http://www.ecy.wa.gov/programs/wq/plants/management/joysmanual/streamtss.html].http://www.ecy.wa.gov/programs/wq/plants/management/joysmanual/streamtss.html accessed 7 may 2014.

21. (2010) Water Market: Mexico Payment for Hydrological Services. Ecosystem Marketplace: A Forest Trends Initiative. The Katoomba Group.

22. Engel S, Pagiola S, Wunder S (2008) Designing payments for environmental services in theory and practice: An overview of the issues. Ecol Econ 65(4):663-674. [http://water.epa.gov/lawsregs/lawsguidance/cwa/tmdl/nutriosocreek.cfm].

23. EPA (2013) Total Maximum Daily Loads (TMDLs) at Work: Arizona. Committed Landowner Implements and Advocates Improved Grazing Practices in Nutrioso Creek, Reducing Turbidity Levels to Attain Water Quality Standards. United States Environmental Protection Agency, . Accessed 5 May 2014.

24. (2007) The State of Food and Agriculture: Paying Farmers for Environmental Services. Food and Agriculture Organization of the United Nations, Rome.

25. Farber S, Costanza R, Childers DL, Erickson J, Gross K, Grove M, Hopkinson CS, Kahn J, Pincetl S, Troy A (2006) Linking ecology and economics for ecosystem management. Bioscience 56(2):121-133

26. Ferreira CEG (2012) Sistema de Suporte a Decisao Espacial Aplicado a Analise da Vulnerabilidade dos Recursos Hidricos na Bacia Guapi-Macacu. Universidade Estadual do Rio de Janeiro, Rio de Janeiro, Brazil.

27. Fidalgo E, Pedreira B, Abreu MD, Moura ID, Godoy M (2008) Uso e Cobertura da Terra da Bacia Hidrografica do rio Guapi-Macacu. Embrapa Solos, Brazil.

28. Forero FA (2002) Sistemas de producción rurales en la Región Andina colombiana: análisis de su viabilidad económica, ambiental y cultural. Grupo Sistemas de Producción y Conservación/ Instituto de Estudios Rurales (IER), Bogota, Colombia.

29. Gaese H (2009) Demands for Interdisciplinary Research in Human-Ecological Systems. In: Gaese H, Torrico JC, Wesenberg J, Schluter S (eds) Biodiversity and Land Use Systems in the Fragmented Mata Atlantica of Rio de Janeiro, Cuvillier Verlag, Gottingen, Germany.

30. Garcia-Nieto AP, Garcia-Llorente M, Iniesta-Arandia I, Martin-Lopez B (2013) Mapping forest ecosystem services: From providing units to beneficiaries. Ecosystem Services 4(Special Issue on Mapping and Modelling Ecosystem Services):126-138

31. Grieg-Gran M, Porras I, Wunder S (2005) How can market mechanisms for forest environmental services help the poor? Preliminary lessons from Latin America. World Dev 33(9):1511-1527

32. Guedes FB, Seehusen SE (2011) Pagamentos por Servicos Ambientais na Mata Atlanica: licoes aprendidas e desafios. Ministerio do Meio Ambiente-MMA, Brasilia, Brazil.

33. Haines-Young R, Potschin M (2013) Proposal for a Common International Classification of Ecosystem Goods and Services (CICES) for Integrated Environmental and Economic Accounting. [http://www.wri.org/publication/forests-water]

34. Hanson C, Talberth J, Yonavjak L (2011) Forests for Water: Exploring Payments for Watershed Services in the U.S. South. World Resources Institute, WRI Issue Brief. Available via WRI. . Accessed 04 Jul 2014. [http://www.environmentalmanager.org/wp-content/uploads/ 2008/ 04/ valuation%2520of%2520ecosystems.pdf]

35. Hawkins K (2003) Economic Valuation of Ecosystem Services. University of Minnesota, . Accessed 04 Jul 2014.

36. Holmes TP (1988) The offsite impact of soil erosion on the water treatment industry. Land Econ 64(4):356-366

37. Honey-Roses J, Acuna V, Bardina M, Brozovi N, Marce R, Munne A, Sabater S, Termes M, Valero F, Vega L, Schneider DW (2013) Examining the Demand fo Ecosystem Services: The Value of

Stream Restoration for Drinking Water Treatment Managers in the Llobregat River, Spain, vol 90:196.205. Elsevier BV

38. (2012) Paying for Watershed Services: An Effective Tool in the Developing World.

39. (2012) Compensation for Hydrological Environmental Services in Los Negros Cloud Forest. Payments for Watershed Markets-Information on Schemes in Developing Countries.

40. Instituto Terra Mater (2009) Relatório da Análise de Percepção Ambiental e sobre Pagamentos por Serviços Ambientais nas Microbacias do Moinho e Cancan. Produto 3. Serviços de monitoramento sócio-econômico e de percepção ambiental em microbacias piloto – "Projeto de pagamento de serviços ambientais". Contrato no PRMC/GEF/BIRD, Piracicaba, Brazil

41. López A (1997) Análise dos Custos Privados e Sociais da erosão do solo- o caso da bacia do Rio Corumbatai. Escola Superior de Agricultura "Luiz de Queiroz" (ESALQ), University of São Paulo (USP), Piracicaba.

42. Lorz C, Abbt-Braun G, Bakker F, Borges P, Bornick H, Frimmel FH, Gaffron A, Hebben N, Hofer R, Makeschin F, Neder K, Roig HL, Steiniger B, Strauch M, Walde D, Weis H, Worch E, Wummel J (2011) Die Bedeutung von Landnutzungsanderungen fur ein Integriertes Wasserressourcen-Management. Eine Fallstudie aus dem westlichen Zentral-Brasilien. . Fachberichte Wasserversorgung. Fachberichte Wasserversorgung, Vulcan Verlag, Essen, Germany.

43. Marques FM (2009) Valoracao dos Servicos Ambientais da Floresta de Mata Atlantica associados a Qualidade e Quantidade da Agua na APA do Sana. Curso de Pos-Graduacao em Agronomia Ciencia do Solo. Seropedica, Universidade Federal Rural do Rio de Janeiro (UFRRJ), Rio de Janeiro, Brazil.

44. Martinez de Anguita P, Rivera S, Beneitez JM, Cruz F, Espinal FM (2011) A GIS cost-benefit analysis-based methodology to establish a payment for environmental services system in watersheds: application to the Calan River in Honduras. J Sustain For 30(1–2):79-110

45. Martin-Ortega J, Ojea E, Roux C (2012) Payments for Water Ecosystem Services in Latin America: Evidence from Reported Experience. TEEB Conference 2012. Mainstreaming the Economics

of Nature: Challenges for Science and Implementation, Leipzig, Germany. ID143

46.   (2010) Programa Nacional de Crédito Fundiário (PNCF). Linha de Financiamento combate à pobreza rural. Ministry of Agrarian Development (MDA). Secretaria de Reordenamento Agrario, Brasilia, Brazil.

47.   (2005) The Millenium Ecosystem Assessment. Island Press, Washington D.C.

48.   Medeiros R, Young CEF, Pavese HB, Araujo FFS (2011) Contribuição das unidades de conservação para a economia nacional: Sumário Executivo. UNEP-WCMC, Brasilia, Brazil.

49.   (2005) Rio Floresta. Assistencia Tecnica e extensao florestal aos agricultores familiares da Mata Atlantica do Estado do Rio de Janeiro. Bases Legais e tecnicas para implantacao de projetos florestais. Ministry of Environment Niteroi, Rio de Janeiro, Brazil. [http://rua.ua.es/dspace/handle/10045/7641?locale=en]

50.   Montes C (2007) Del desarrollo sostenible al servicio de los ecosistemas. Asociacion Espanola de Ecologia Terrestre. Ecosistemas Revista Cientifica y Tecnica de Ecologia y Medio Ambiente 16(3). Available via RUA.. Accessed 04 Jul 2014.

51.   Moore W, McCarl B (1987) Off-site costs of soil erosion: A case study in the Willamette Valley. West J Agric Econ 12(1):42-49

52.   Muñoz-Piña C, Guevara A, Torres JM, Braña J (2008) Paying for the hydrological services of Mexico's forests: Analysis, negotiations and results. Ecol Econ 65(4):725-736.[http:/ / dinario.fh-koeln. de/ 2011/ F.%20Naegeli-%20Evaluation%20of%20f orest%20 fragmentation%20and%20land% 20use.pdf]

53.   Naegeli F (2010) Evaluation of Forest Fragmentation and Land Use Change Patterns using Remote Sensing Techniques and Field Methods, Master Thesis. Cologne University of Applied Sciences, Abstract available via DINARI. Accessed 04 Jul 2014

54.   Nehren U, Kirchner A, Sattler D, Turetta AP, Heinrich J (2013) Impact of natural climate change and historical land use on landscape development in the Atlantic Forest of Rio de Janeiro, Brazil. An Acad Bras Cienc 85(2):497-518.[http://postconflict. unep.ch/publications/DRR_CASE_STUDIES_&_EXERCISES.pdf]

55. Nehren U, Sudmeier-Rieux K, Sandholz S, Estrella M, Lomarda M, Guillén T (eds) (2014) The Ecosystem-based Disaster Risk Reduction Case Study and Exercise Source Book. UNEP/CNRD, . Accessed 04 Jul 2014

56. Pabon-Zamora L, Fauzi A, Halim A, Bezaury-Creel J, Vega-Lopez E, Leon F, Gil L, Cartaya V (2008) Protected Areas and Human Well-Being: Experiences from Indonesia, Mexico, Peru and Venezuela. Secretariat of Convention on Biological Diversity, Montreal.

57. Pagiola S, von Ritter K, Bishop J (2004) Assessing the Economic Value of Ecosystem Conservation. The World Bank Environment Department, Washington, United States of America.

58. Pagiola S, Zhang W, Colom A (2010) Can payments for watershed services help finance biodiversity conservation? A spatial analysis of Highland Guatemala. J Nat Resour Pol Res 2(1):7-24

59. Paiva M, Penedo S, Kuenne A, Prado RB, Schuler AE (2011) Qualidade da Agua e Exportacao de Sedimentos em Sub-bacias dos rios Guapi-Macacu- Bioma Mata Atlantica- RJ. In: XXXIII Congresso Brasileiro de Ciencia do Solo. Solos nos Biomas Brasileiros: Sustentabilidade e Mudancas Climaticas. Embrapa, Uberlandia /Minas Gerais, Brazil. [http://ainfo.cnptia.embrapa.br/digital/bitstream/item/51926/1/2315-1.pdf].Available via Embrapa. http://ainfo.cnptia.embrapa.br/digital/bitstream/item/51926/1/2315-1.pdf. Accessed 04 Jul 2014.

60. Pattanayak SK, Wunder S, Ferraro PJ (2010) Show me the money: Do payments supply environmental services in developing countries? Rev Environ Econ Policy 4(2):254-274.

61. Pedreira B, Fidalgo ECC, Abreu MB, Epiphanio JCN, Galvao LS (2009) Mapeamento do uso e cobertura da terra da bacia hidrografica do rio Guapi-Macacu, RJ. Anais XIV Simposio Brasileiro de Sensoriamento Remoto. Natal, INPE. pp 2111–2118.[http:/ / dutraeconomicus.files.wordpress.com / 2014/ 01/ roger-perman-yue-ma-michael-common- david-maddison-james-mcgilvray-natu ral-resource-and-environmental-econ omics-3rd-edition-2003.pdf]

62. Perman R, Ma Y, McGilvray J, Common M (2003) Natural Resource and Environmental Economics, Third Editionth edn. Pearson Education Limited, Available via WordPress. . Accessed 04 Jul 2014.[http://pubs.iied.org/13542IIED.html]

63. Porras I, Grieg-Gran M, Neves N (2008) All that Glitters: A Review of Payments for Watershed Services in Developing Countries, vol Natural Resource Issues 11. International Institute for Environment and Development, Available via IIED. . Accessed 04 Jul 2014.

64. Postel B, Barton H, Jr T (2005) Watershed protection: Capturing the benefits of nature's water supply services. Nat Res Forum 29:98-108

65. Pria DA, Diederichsen A, Klemz C (2013) Pagamento por Servicos Ambientais. Uma estrategia para a conservacao ambiental nas regioes produtivas do Brasil? Sustentabilidade em. Debate 4(1):317.340 [http:/ / www.google.de/ url?url=http:/ / comunidadpmpca.uaslp.mx/ documento. aspx%3FidT%3D258&rct=j&fr m=1&q=&esrc=s&sa=U&ei=z9 q2U8KDLOTmy wPbw4CgCA&ved=0CCQQFjAD&usg=AFQjC NF ldcTMmPE0Err5jbaXkOJOA3FeJA]

66. Quintana B (2012) Native Tree Species in Silvopastoral Systems: A Bioeconomic Assessment in Cachoeiras de Macacu, RJ-Brazil, Master Thesis. Master of Science awarded by Universidad Autónoma de San Luis Potosí and Cologne University of Applied Sciences. Cologne University of Applied Sciences. Available. . Accessed 04 Jul 2014

67. Quintero M, Wunder S, Estrada R (2009) For services rendered? Modeling hydrology and livelihoods in Andean payments for environmental services schemes. For Ecol Manag 258(9):1871-1880

68. Reis LVS (2004) Cobertura Florestal e Custo do Tratamento de Águas em Bacias Hidrográficas de Abastecimento Público: Caso do Manancial do Município de Piracicaba. University of São Paulo (USP), Piracicaba.

69. Ribeiro MC, Metzger JP, Martensen AC, Ponzoni FJ, Hirota MM (2009) The Brazilian Atlantic Forest: How much is left, and how is the remaining forest distributed? Implications for conservation. Biol Conserv 142(6):1141-1153

70. (2013) Rio Rural. Programa de Desenvolvimento Rural Sustentavel em Microbacias Hidrograficas. Governo do Rio de Janeiro. Secretaria de Agricultura e Pecuaria, Rio de Janeiro, Brazil.

71. Rodriguez Osuna V (2013) Smallholder Production and Climate Risk: The Lower Amazon Region, Brazil. LAP Lambert Academic

Publishing, Germany.[http:/ / www.teebweb.org/ publication/ the-economics-of-ecosystems-and-bio diversity-teeb-for-water-and-wetlan ds/]

72. Russi D, ten Brink P, Farmer A, Badura T, Coates D, Forster J, Kumar R, Davidson N (2013) The Economics of Ecosystems and Biodiversity for Water and Wetlands. Institute for European Environmental Policy (IEEP) & Ramsar Secretariat. Available via TEEB. . Accessed 04 Jul 2014

73. Santos DG, Domingues AF, Gisler CVT (2010) Gestão de recursos hídricos na agricultura: O Programa Produtor de Água. In: Prado RB, Turetta AP, Andrade AG (eds) Manejo e Conservação do Solo e da Água no Contexto das Mudanças Ambientais, Brazil, Rio de Janeiro. pp 353-376

74. (2011) Government of Rio de Janeiro. Campanha de Vacinalção contra a febre aftosa. Cachoeiras de Macacu Region. Núcleo de Defesa Sanitária- Escritorio Cachoeiras Macacu, Rio de Janeiro, Brazil.

75. Stanton T, Echavarria M, Hamilton K, Ott C (2010) State of Watershed Payments. An Emerging Marketplace. Executive Summary. Ecosystem Marketplace: A Forest Trends Initiative.

76. Strobel JS, De Sousa JR WC, da Motta RS, Amend MR, Goncalves DA (2007) Critérios Econômicos para a Aplicação do Princípio do Protetor – Recebedor: Estudo de Caso do Parque Estadual dos Três Picos Serie Tecnica. Conservation Strategy Fund, Brazil.

77. (2009) The Economics of Ecosystems and Biodiversity for National and International Policy Maker. Summary: Responding to the Value of Nature. The Economics of Ecosystems and Biodiversity (TEEB), Wesseling, Germany.

78. (2010) The Economics of Ecosystems and Biodiversity (TEEB). Ecological and Economic Foundations, Earthscan, London and Washington.

79. (2010) Mainstreaming the Economics of Nature: A Synthesis of the Approach, Conclusions and Recommendations of TEEB. The Economics of Ecosystems and Biodiversity (TEEB), Malta.

80. (2010) TEEB for Business Report. The Economics of Ecosystems and Biodiversity (TEEB), Malta.

81. Thomas JM, Callan SJ (2010) Economia Ambiental: Aplicacoes, Politicas e Teoria. Cencage Learning, São Paulo.

82. van Noordwijk M (2005) RUPES typology of environmental service worthy of reward. ICRAF-Southeast Asia, Bogor.

83. Veiga FC (2008) A Construcao dos Mercados de Servicos Ambientais e suas Implicacoes para o Desenvolvimento Sustentavel no Brasil. UFRRJ, Rio de Janeiro, Brazil.

84. Veiga FC, Galvadao M (2011) Iniciativas de PSA de conservacao dos recursos hidricos na Mata Atlantica. In: Guedes BF, Seehusen ES (eds) Pagamentos por Servicos Ambientais na Mata Atlanica: licoes aprendidas e desafios, d Ministerio do Meio Ambiente, Secretaria de Biodiversidade e Florestas, Departamento de Conservacao da Biodiversidade, Brasilia, Brazil.

85. (2011) Estimating the opportunity costs of REDD+. The World Bank Institute, Washington DC.

86. (2011) Mainstreaming Ecosystem Services Initiative (MESI). World Resources Institute, Washington, United States of America.

87. Wunder S, Alban M (2008) Decentralized payments for environmental services: The cases of Pimampiro and PROFAFOR in Ecuador. Ecol Econ 65(4):685-698

88. Zimmer Y, Deblitz C, Seifert K (2009) Die globale Landwirtschaft besser verstehen : Agrarokonomen betreiben internationales Netzwerk agri benchmark. Agribenchmark, Braunschweig, Germany.

Chapter 7

Potential Water-related Environmental Risks of Hydraulic Fracturing Employed in Exploration and Exploitation of Unconventional Natural Gas Reservoirs in Germany

Axel Bergmann[1], Frank-Andreas Weber[1],
H Georg Meiners[2], and Frank Müller[2]

[1]IWW Water Centre, Department Water Resources Management, Moritzstrasse 26, Muelheim 45476, Germany
[2]ahu AG Wasser Boden Geomatik, Kirberichshofer Weg 6, Aachen 52066, Germany

ABSTRACT

Background

The application of hydraulic fracturing during exploration and exploitation of unconventional natural gas reservoirs is currently under intense public discussion. On behalf of the German Federal Environment Agency we have investigated the potential water-related environmental risks for human health and the environment that could be caused by employing hydraulic fracturing in unconventional gas reservoirs in Germany. Here we provide an overview of the present situation and the state of the debate in Germany and summarize main results of the conducted risk assessment.

Results

We propose a concept for a risk assessment considering the site-specific analysis of the geosystem, the relevance of possible impact pathways and the hazard potential of the fracking fluids employed. The foundation of a sound risk analysis is a description of the current system, the relevant impact pathways and their interactions. An evaluation of fracking fluids used in Germany shows that several additives were employed even in newer fluids that exhibit critical properties or for which an assessment of their behaviour and effects in the environment is not possible or limited due to lack of current knowledge. The authors propose an assessment method that allows for the estimation of the hazard potential of specific fracking fluids, formation water, and the flowback based on legal thresholds and guidance values as well as on human- and eco-toxicologically predicted no-effect concentrations. The assessment of a previously employed and a prospectively planed fracking fluids shows that these fluids exhibit a high hazard potential. The flowback containing fracking fluid, formation water, and possibly reaction products can also exhibit serious hazard potentials, requiring environmentally acceptable techniques for its treatment and disposal.

Conclusions

The risk analysis must be conducted always site-specifically and consider regional groundwater flow conditions. The study concludes that currently missing knowledge and data prevent a profound assessment of the risks and their technical controllability in Germany. Missing knowledge and information includes data on the properties of the deep geosystem and of the behaviour and effects of the deployed chemical additives. In this setting the authors propose several recommendations for further action and procedures regarding the application of hydraulic fracturing in unconventional gas reservoirs in Germany.

BACKGROUND

The application of hydraulic fracturing ("fracking") in the exploration and exploitation of unconventional natural gas reservoirs has been generating intensive public debates in a variety of countries. Major concerns have focused on the potential impacts, hydraulic fracturing may cause on the environment and on human health, especially if fracking fluids contain toxic and environmentally harmful chemical additives.

Unconventional gas reservoirs are proven or presumed to be present in a number of different geological formations. An overview of potential geological host formations of unconventional gas reservoirs in Germany is given in Table 1, differentiating coalbed methane (CBM), shale gas and tight gas reservoirs. According to current estimates [1], the technologically recoverable gas reserves present in shale gas reservoirs in Germany amount to about 1,300 billion m^3 (estimates range from 0.7 to 2.3 $\cdot$ 10^{12} m^3), assuming that 10% of the gas in place (GIP) is technologically recoverable. This estimated range of technologically recoverable shale gas reservoirs could, if exploited completely, cover the current annual gas consumption of Germany for 8 to 27 years [2]. The GIP in coalbed methane reservoirs in Germany is estimated to 450 billion m^3[3], but the technologically recoverable fraction has not yet been analysed. Conventional gas and tight gas reservoirs have been exploited in Germany over several decades, but current estimates of GIP remaining (100 billion m^3 and 20 billion m^3, respectively [3]) indicate that the remaining reserves are limited.

Table 1: Potential unconventional gas reservoirs in Germany

Type of reservoir	Most promising reservoir	Regions
Coal bed methane (source rocks)	Seam-bearing Upper Carboniferous	Northern Ruhr region/ Münsterland Basin (NRW)
		Ibbenbühen (NRW)
		Saar Basin (Saarland)
Shale gas (source rocks)	Tertiary clay formations (e.g. Fischschiefer)	Molasse Basin (BW)
	Posidonia Shale (Black Jurassic)*	Northwest German Basin (e.g. Lünne) (NI)
	Wealden clay formations (e.g. Lower Cretaceous)*	Molasse Basin (BW) Upper Rhine Graben
	Permian clay formations (e.g. black shale (stinkschiefe"), copper shale)	Weser Depression (NRW/ NI) Northeast German Basin (NI/SA)
	Carboniferous and Devonian clay formations e.g. alum shale (Lower Carboniferous)*	Northern edge of the Rhenish massif (NRW) Northwest German Basin
	Silurian slates	Harz Mountain (NI/SA)
	Cambro Ordovician clay formations ("alum shale")	Northeast German Basin (not yet studied in details)
Tight gas (deposit rocks)	Red sandstone Permian sandstones (Rotliegend) and carbonates (Zechstein)	Northwest German basin (NI/SA) Northeast German basin (e.g. Leer) (NI)
	Permian sandstones (Rotliegend) and dolomite (Stassfurt series) sandstones (Triassic)	Thuringian Basin (TH) Northwest German Basin (e.g. Vechta) (NI)
	Upper Carboniferous sandstones	

*indicates most relevant shale gas reservoirs according to[1].
Bergmann et al.

Bergmann et al. Environmental Sciences Europe 2014 26:10
doi:10.1186/2190-4715-26-10

The mining authorizations that have been issued for the exploration of unconventional gas reservoirs in Germany are shown in Figure 1. Most exploration has yet focused on the recovery and analysis of drilling core material as well as on geophysical methods, but hydraulic fracturing has already been applied in exploration at two sites [4]: at the site Damme 3 in Lower Saxony (3 fracs in the Wealden clay formation in depth of 1,045 – 1,530 m below ground surface using a slickwater fracking fluid in 2008) and at the site Natarp in North Rhine-Westphalia (2 fracs in CBM reservoirs in depth of 1,800 – 1,947 m using a gel fluid in 1995). To our knowledge, no mining authorizations have yet been approved for production-oriented exploitation of shale gas or CBM reservoirs in Germany. In the ongoing exploitation of tight gas and conventional gas reservoirs, however, experience in using hydraulic fracturing has been gained by pumping over 300 fracs over the last decades, mainly in the federal state of Lower Saxony [4]. In general, the exploited tight gas reservoirs in Germany are located in greater depth (often > 3.500 m) than some of the shale gas and CBM reservoirs currently considered for exploration, which vary in depth but are located partly in depth of 1.000 m or less[2,4-6], raising additional concerns on potential impacts on near-surface groundwater resources.

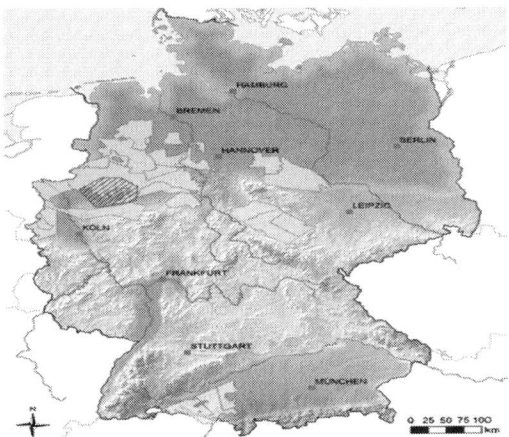

Figure 1: Mining authorizations in Germany (yellow, last revision: 31 December 2011) for exploration for unconventional hydrocarbon reservoirs (ochre: regions with the basic geological conditions for formation of shale gas) [1].

Driven by reports on the application and risk assessment of hydraulic fracturing in the U.S. [7-11], several risk assessments have recently been conducted on the specific German geological, technical, and legal situation, including an investigation on behalf of the German Federal Environment Agency (UBA) [4], a survey on behalf of the Ministry for Climate Protection, Environment, Agriculture, Nature Conservation and Consumer Protection (MKULNV) of the federal state of North Rhine-Westphalia [5], and an investigation of an independent expert group initiated by ExxonMobil Production Germany GmbH [12]. Given the current state of exploration of shale gas and CBM reservoirs in Germany, most risk assessments were conducted generically (i.e. not site-specific) or focused on some selected geological settings. Two site-specific investigations on regional situations in northern Hessian and in the river Ruhr watershed have recently been conducted [6,13].

Current State of the Debate in Germany

The political debate on hydraulic fracturing in Germany has proceeded as a result of the conducted risk assessments (or independent thereof), and new administrative procedures have been adapted recently.

The State Authority for Mining, Energy and Geology (LBEG) of Lower Saxony has issued minimum requirements for operating plans, criteria, and approval procedure for hydraulic treatments of boreholes in petroleum and natural gas reservoirs [14]. ExxonMobil Production Germany GmbH, a major operator in Germany, has announced that fracking projects in the vicinity of certain mineral spa protection zones are not further pursuit and no further hydraulic fracturing activities are carried out before suitable concepts for groundwater monitoring are implemented [15].

The state of North Rhine-Westphalia is currently not approving any exploration or production of natural gas from unconventional gas reservoirs, if harmful substances are employed [16]. A dialogue process is planned to involve the gas industry and communities, citizens, and relevant institutions in developing criteria for project approval and eliminating deficits of information and knowledge. In this context, borehole investigations, excluding hydraulic fracturing, are discussed for research purposes[17].

According to current press communications [18], the state of Lower Saxony is not approving further exploration and exploitation of shale gas and CBM reservoirs based on the lack of adequate risk assessment, but plans to continue approving exploitation of tight gas reservoirs in sandstone formation in depths > 2.500 m, as long as no environmentally toxic substances are injected underground.

Draft legislations amending the environmental impact assessment (EIA) regulation and of the Water Management Act (WHG) are currently discussed in Germany [19]. The drafts call for a ban of deep drillings involving hydraulic fracturing and the underground disposal of flowback in water protection zones, mineral spa protection zones, and in catchment areas of natural lakes from which raw water is procured directly for the public water supply. Based on the discussed draft legislation, the catchment area of artificial lakes and dams from which water is indirectly obtained for drinking water purposes would not generally be considered an exclusion zone [20].

Two regional investigations have analysed the regional occurrence of shale gas reservoirs in comparison to competing land-use obligations [6,13]. In a study on behalf of the river Ruhr water works consortium (Arbeitsgemeinschaft der Wasserwerke an der Ruhr e.V.) and the Ruhr River water board (Ruhrverband), we concluded that considering the regional occurrence of the shale gas reservoirs, the exclusion areas proposed by the draft legislations, and adopting criteria for the approval of exploitation involving hydraulic fracturing issued in Lower Saxony, an area of less than 3% of the issued mining authorization is accessible for exploitation of the shale gas reservoirs. Furthermore, a legal expertise commissioned by the Hessian Ministry of Environment, Energy, Agriculture and Consumer Protection (HLUG) has noted [13] that mining authorizations must not be granted if competing obligations among public stakeholders preclude subsequent exploitation of the gas reservoirs in the entire allocated field.

In the so-called "Hannover-Erklärung", the Federal Institute for Geosciences and Natural Resources (BGR), the Helmholtz Centre Potsdam - GFZ German Research Centre for Geosciences and the Helmholtz Centre for Environmental Research (UFZ) have called for the development of environmentally friendly fracking technology and proposed joint demonstration projects involving industry, research institutions, environmental organizations, and the general public [21].

An alliance of water suppliers, the Ruhr River water works consortium, and members of the beverage industry have called for clear legal provisions to protect the safety and purity of water resources from impacts of hydraulic fracturing in the so-called "Gelsenkirchener Erklärung" [22].

Furthermore, the exploitation of shale gas reservoirs is currently discussed controversial from an energy policy point of view. While the Federal Institute for Geosciences and Natural Resources (BGR) concluded that shale gas can contribute to domestic energy security [1], the German Advisory Council on the Environment (SRU) comes to the conclusion that the exploitation of shale gas using hydraulic fracturing is not necessary in Germany from an energy policy point of view and cannot substantially contribute to the transition to renewable energy sources [2].

Objectives

On behalf of the German Federal Environment Agency (UBA), a consortium of IWW Water Centre, ahu AG, [Gaßner, Groth, Siederer & Coll.], and Technical University of Darmstadt, has conducted a comprehensive investigation on potential environmental impacts of hydraulic fracturing related to exploration and exploitation of unconventional natural gas reservoirs, which focused on the framework of a risk assessment, the analysis of potential impact pathways, a method for assessing the hazard potentials of the fracking fluids employed, and on legal regulations and administrative structures [4]. Here we summarize main results of this study and propose recommendations for action and procedures. The study is based mostly on publicly accessible information including the relevant literature available internationally, but also on information provided by German authorities and operating companies.

RESULTS AND DISCUSSION

For assessing the risks that the application of hydraulic fracturing in unconventional natural gas reservoirs can pose on the water environment, we propose a concept that considers both the possible impact pathways and the potential hazard, any migration of the substances employed or encountered along these impact pathways could cause on exploitable water resources (Figure 2). Only if impact pathways are relevant for substance migration on the time scale considered, the substance-related hazard potentials cause adverse effects on the exploitable water environment. The risk of contamination of exploitable water resources is thus obtained by multiplying the relevance of the impact pathway(s) and the hazard potential of the pertinent fluids (fracking fluids and formation water). Since the state of knowledge does currently not allow for numerical calculations, we propose a five-part scale to evaluate the relevance of impact pathways and the hazard potential of the fluids involved (Figure 2).

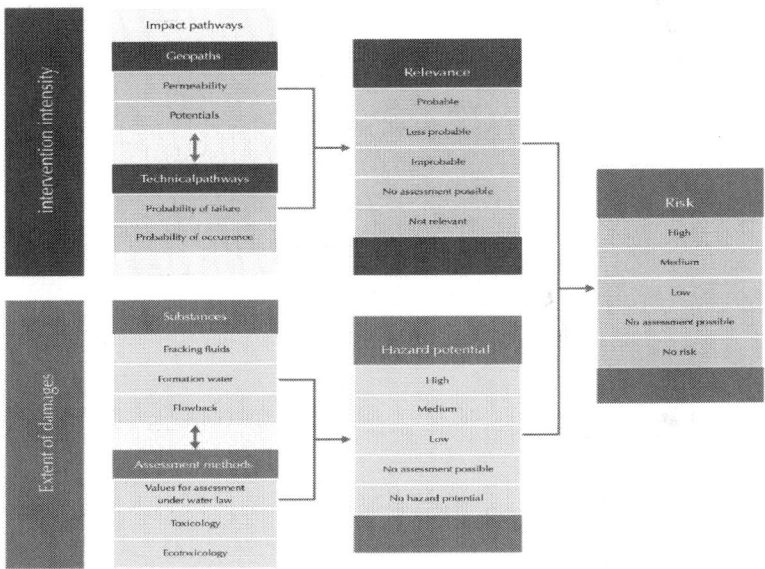

Figure 2: Structure of risk analysis for assessment of unconventional gas exploitation.

Impact Pathways

Potential water-related impact pathways are shown schematically in Figure 3, considering both technical and geological impact pathways. In most cases, failures of technical systems need to occur (such as failures of the well casing) for activating potential geological impact pathways (such as migration along faults), except in the fracking horizon, where no technical barrier is in place. Technical impact pathways could be quantified by probabilities of occurrence or probabilities of failure if data suitable for the German geological, technical and legal conditions were available. For a geological impact pathway to be relevant for substance migration, both permeability and hydraulic potentials must be considered for each geosystem site-specifically. Without suitable numerical quantification, however, the relevance of geological impact pathways can be estimated only with great uncertainties, for example using worst-case approaches.

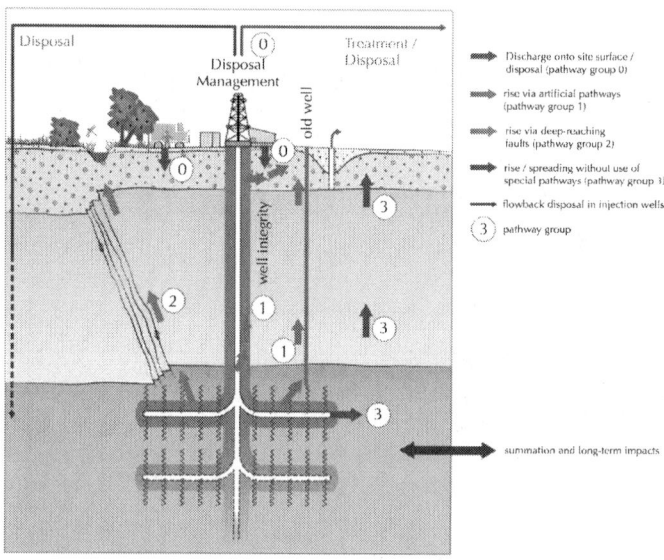

Figure 3: Schematics of potential impact pathways.

Pathway group 0 refers to (pollutant) discharges that occur directly at the ground surface, and especially in handling of fracking fluids (transport, storage, etc.) or flowback (e.g. via accidents or improper handling).

Pathway group 1 refers to potential (pollutant) discharges and migration along wells, i.e. to artificial underground pathways. With regard to the impact pathways involved, a distinction has to be made between production wells and old wells, such as wells from other explorations and uses.

Pathway group 2 comprises all impact pathways along geological faults. Significantly, the permeability along any given fault can vary, section-wise. Whereas deep-reaching, continuous faults can often be monitored, since the near-surface locations of their outcrops are usually known, faults that affect only parts of the overburden are difficult to monitor.

Pathway group 3 comprises extensive rise, as well as lateral spreading, of gases and fluids through geological strata (for example, via an aquifer), without preferred pathways similar to those described for pathway groups 1 and 2. Impact pathways in pathway group 3 depend primarily on the prevailing geological and hydrogeological conditions.

Summation and combination effects of the aforementioned impact pathways must be taken into account appropriately. Since many flow processes in the deep underground take place slowly, the relevant long-term impacts need to be considered. Such estimation is possible only on the basis of an extensive understanding of the geological and hydrogeological conditions prevailing in deep underground horizons, although not enough data of the studied geosystems are currently available to support conceptual or even numerical models.

Furthermore, the flowback disposal needs to be assessed as additional impact pathway, especially if flowback disposal is via injection into underground disposal wells.

Fracking Fluids

Overview

The fracking fluid is the hydraulic medium used for applying pressure to the rock strata inducing fracturing. With the fracking fluid, proppants (such as quartz sand) are transported into the created fractures in order to keep fractures from closing under the pressure of the surrounding rock and, thus, to ensure that the pathways created remain accessible for gas migration during the production phase. Fracking fluids usually contain a variety of chemical additives, with functions such as facilitating transport of proppants into fractures, preventing formation of precipitates, microbiological growth, formation of hydrogen sulphide, swelling of clay minerals, corrosion, and reducing fluid friction at high pump rates. Table 2 provides an overview of the functions of certain additives.

Table 2: Functions of additives used in fracking fluids (based on [4],[9])

Additive	Function
Proppants	Keeping the fractures created open under the pressure of the surrounding rock and allows gas/fluid to flow to the well bore
Scale inhibitors	Preventing deposits of poorly soluble precipitates, such as carbonates and sulphates
Biocides	Preventing bacterial growth, biofilm formation and formation of hydrogen sulphide by sulphate-reducing bacteria
Iron control	Preventing iron-oxide precipitation
Gelling agents	Improving proppant transport
High-temperature stabilizer (temperature stabilizer)	Preventing gel decomposition at high temperatures within the target horizon
Breakers	Reducing the viscosity of gel-containing fracking fluids for depositing proppants
Corrosion inhibitors	Protecting against equipment corrosion
Solvents	Improving the solubility of additives
pH regulators and buffers (pH control)	Controlling the pH of tracking fluids
Crosslinkers	Increasing viscosity at higher temperatures, to improve proppant transport
Friction reducers	Reducing friction within frac king fluids
Acids	Pretreating perforated sections of the well, and cleaning them of cement and drilling mud; dissolving acid-soluble minerals
Foams	Supporting proppant transport
H2S scavengers	Removing toxic hydrogen sulphide to protect equipment against corrosion
Surfactants	Reducing surface tension of fluids
Clay stabilizers	Reducing swelling and migration of clay minerals

Bergmann et al.

Bergmann et al. Environmental Sciences Europe 2014 26:10 doi:10.1186/2190-4715-26-10

In the following we present information on the fracking fluids and additives that have so far been employed in Germany. We then presented a method for assessing the hazard potentials of the fracking fluids employed with regard to groundwater, especially with regard to human use of groundwater as drinking water, and as part of natural cycles. In applying the method we assess selected fracking fluids used in Germany to date and possible new improvements of such fluids.

Fracking Fluids Used in Germany

We relied primarily on publicly accessible data to obtain information on the fracking fluids used in unconventional reservoirs in Germany [23]; only in some cases information from non-publicly accessible sources were obtainable [24]. The information on the composition of the fracking fluids used is based mainly on analyses of safety data sheets of the commercial products used to prepare fracking fluids. It has been found that these safety data sheets are often the only available source of information on the identity and the concentrations of the additives used. For approval authorities, this situation creates considerable uncertainties and lack of knowledge regarding the identity and the quantities of additives actually injected into the borehole.

Quantities Used

Information on fluid volumes was available for a total of 30 fracking fluids used in various unconventional reservoirs (and in one conventional reservoir) in Germany between 1982 and 2011. Most of the reservoirs in which the fluids were injected were tight gas reservoirs in Lower Saxony. The quantities used varied considerably, depending on the type of fracking fluid and the characteristics of the reservoirs. The quantities of fracking fluids used per frac ranged from less than 100 m^3 to more than 4,000 m^3. With the modern gel fluids used since 2000, an average of about 100 t of proppants and about 7.3 t of additives (of which usually less than 30 kg were biocides) were injected per frac. The quantities used can be quite large especially with multi-frac stimulations and/or use of slickwater fluids: for example, a total of about 12,000 m³ of water, 588 t of proppants, and 20 t of additives (of which 460 kg were biocides) were injected into the "Damme 3" borehole in three frac operations in 2008.

Commercial Hydraulic Fracturing Products

According to the available information, at least 88 different hydraulic fracturing products have been used to prepare fracking fluids in Germany. However, since data are available on only 21 fracking fluids (corresponding to about 21% of the approximate 300 fracs carried out in Germany), it must be assumed that other products have also been employed. For 80 of the 88 products, we were able to obtain manufacturers' or importers' safety data sheets that were either current or valid at the time the fracs were carried out. Evaluation of the available 80 safety data sheets revealed that:

- 6 products are classified as toxic,
- 6 are classified as dangerous to the environment,
- 25 are classified as harmful,
- 14 are classified as irritant,
- 12 are classified as corrosive, and
- 27 are classified as non-hazardous

according to directives 67/548/EEC or 1999/45/EC, respectively. Several products are classified in more than one hazard class. With respect to the German water hazard classification (Wassergefährdungsklasse WGK), the commercial products were classified as follows according to the information in the safety data sheets:

- 3 preparations are classified as severely hazardous to waters,
- 12 preparations are classified as hazardous to waters,
- 22 preparations are classified as low hazardous to waters,
- 10 preparations are classified as not hazardous for water.

A total of 33 of the safety data sheets available to the study authors provided no information on the water hazard class of the product.

Fracking Additives

Information on the fracking additives used in the hydraulic fracturing products was available to the study authors for 28 fracking fluids. Those fluids were used in about 25% of 300 fracs carried out in Germany. Evaluation of those 28 fracking fluids showed that, overall, at least 112 substances/substance mixtures have so far been used in Germany. For 76 of the 112 substances/substance mixtures, either unique Chemical Abstracts Service (CAS) numbers were provided or it proved possible to correct or determine the CAS number on the basis of a unique given substance name. A total of 36 substances/substance mixtures could not be uniquely identified, either because their composition was unknown or because the available safety data sheets referred only to unspecific chemical group names (such as aromatic ketones, inorganic salts, etc.).

Hazard Potentials of Fracking Fluids

Comparison of Two Fracking Fluids

Since recipes for fracking fluids are normally tailored to specific reservoirs, the hazard potentials of each fluid need to be assessed site-specifically. Based on the assessment method described in the Methods section, we have assessed the two fluids used to date in shale gas and CBM reservoirs in Germany as two examples. Planned improvements of fracking fluids were taken into account by assessing two fluids mentioned by an operator as potentially being suitable for shale gas reservoirs and, possibly, CBM reservoirs (improvements of slickwater and gel fluids) [4].

The hazard potentials of the slickwater fluid employed in the shale gas reservoir in 2008 and a planned improved composition are compared in Table 3. The assessment concludes that the slickwater fluid used in 2008 has a high toxicological and ecotoxicological hazard potential. In the improved fracking fluid, three hazardous additives that were still being used in 2008 are replaced by substances with considerably lower hazard potentials. However, also the improved fluid seems to exhibit a high hazard potential, because of employing high concentrations of a formaldehyde-forming biocide, for which little data is available for assessing its behaviour, fate, toxicity, and formation of degradation products. The replacement of the three hazardous additives that were still being used in 2008 by substances with considerably lower hazard potentials must be critically evaluated, since the underlying database for assessing those additives has been available for years, suggesting that service companies, operators, and/or authorities in the past have not always adequately considered the possibilities of substituting hazardous additives.

Table 3: Composition and hazard potential of two slickwater fluids

Fracking fluid used at Damme 3					Planned improvement of a slickwater fluid			
Function	Additive	Dissolved concentration in fracking fluid	Risk quotient based on toxicological assessment	Risk quotient based on eco-toxicological assessment	Additive	Dissolved concentration in fracking fluid	Risk quotient based on toxicological assessment	Risk quotient based on eco-toxicological assessment
Clay stabilizer	Tetramethyl-ammonium chloride	520 mg/L	1,733,000	Database insufficient (>2,600,000)	Cholinium chloride	750 mg/L	< 43	210
Friction reducer	Hydrotreated light petroleum distillates	220 mg/L	2,200	55,000	Butyl diglycol	350 mg/L	40	6,600
Surfactant	Ethoxylated octylphenol	36 mg/L	120,000	20,000	Polyethylene glycol monohexyl ether	130 mg/L	743	760
Biozide	Isothiazolinone derivative	4 mg/L	7,520	72,000	(Ethylenedioxy)–dimethanol	1,000 mg/L	10,000,000	Database insufficient (139,000)

Assessment of the fracking fluid used 2008 for hydraulic fracturing in a shale gas reservoir at Damme 3 and of a planned improvement based on human- and ecotoxicologically derived risk quotients.

Bergmann et al.

Bergmann et al. Environmental Sciences Europe 2014 26:10
doi:10.1186/2190-4715-26-10

Current developments aiming at reducing the numbers of additives used, at finding substitutes for substances that are highly toxic, carcinogenic, mutagenic, or toxic for reproduction, and at reducing or replacing biocidal agents, point to potential progress in the development of environmentally compatible fracking fluids. However, the authors can currently not evaluate the feasibility or progress of such efforts.

Flowback

Quantities and Composition

After pressure has been applied to the gas-bearing formation, some of the injected fracking fluids are recovered along with formation water and gas extracted from the well. The so-called flowback consists of varying proportions of injected fracking fluids and co-extracted formation water. Initially, fracking fluids account for the larger share of flowback; later, formation water predominates. As a result of various hydrogeochemical processes that can occur within the reservoir horizon (Figure 4), flowback can contain other substances in addition to fracking additives and formation water constituents.

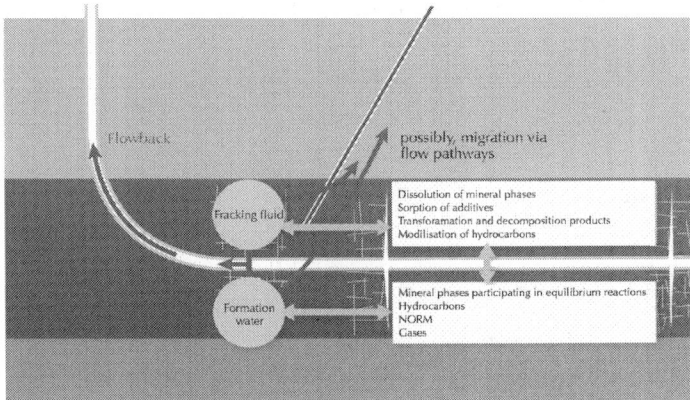

Figure 4: Hydrogeochemical processes affecting flowback formation via mixing of fracking fluids and formation water.

At the high pressures and temperatures prevailing in the target horizon, injected fracking additives may undergo chemical transformation and decomposition reactions in the presence of saline formation water. Microbiological decomposition reactions may occur as soon as the effects of the injected biocides diminish. In the process, metabolites can form that can pose toxicological and ecotoxicological risks potentially even exceeding the hazard posed by the parent substances that were injected.

Because the characteristics of formation water are always reservoir-specific, and because the proportions of extracted fracking additives vary, the characteristics of flowback have to be individually assessed for each site and pertinent time. Little information is available about the constituents of formation water in shale gas and CBM reservoirs in Germany, such as information about primary, secondary, and trace components, dissolved gases, organic substances, and NORM (Naturally Occurring Radioactive Material); regional and depth-oriented data is often missing.

At present, there is a lack of reliable analyses and mass balances that would allow for quantification of the variable mixing fractions, the fraction of the extracted fracking fluid, and possible reaction products. To date, no systematic measurements have been carried out for the purpose of identifying transformation and decomposition products in the flowback. Assessments of flowback from the "Damme 3" borehole carried out by Rosenwinkel et al. [25] concluded that only 8% of injected fracking fluids were being recovered as part of the flowback. Even though that percentage can be expected to increase as production continues, it seems certain that a substantial proportion of the fracking additives injected remains underground.

Disposal of Flowback

Possible technical processes for treating flowback have been reviewed by Rosenwinkel et al. [25] concluding that none of those treatment options, at present, qualifies as "best available technology" within the meaning of the German Federal Water Act. In general, the following options are possibly suitable for disposing or recycling of flowback in Germany:

- Underground injection via disposal wells,

- treatment for discharge into surface water,
- treatment for discharge into the sewer system,
- recycle and reuse in future hydraulic fracturing operations.

Operators currently refer to underground disposal of flowback as an important prerequisite for (cost-effective) exploitation of unconventional gas reservoirs. From the perspective of the study authors, flowback disposal via deep-underground injection can entail risks requiring site-specific risk assessment and monitoring.

CONCLUSIONS

There is general lack of basic information that would be needed for any well-founded assessment of the pertinent risks and the degree to which they can be controlled by technical means. Examples of such missing data include information regarding the structures and properties of deep geological systems (permeabilities, potential differences), the identities of the fracking additives used, and the chemical and toxicological properties of such additives. There are several reasons for this lack of information and data: (a) the information and data are not (openly) accessible, (b) the information and data have not yet been evaluated, and/or (c) there are gaps in knowledge that can only be closed through additional studies and research.

By studying selected geological systems in which shale gas or CBM reservoirs in Germany are found or assumed [4] we concluded that site-specifically certain impact pathways could be relevant for fluid migration. Little reliable data are currently available that would provide a basis for the reliable exclusion of risks to near-surface water resources. Assessment of selected fracking fluids used in unconventional gas reservoirs in Germany, along with the available information on the characteristics of flowback, have revealed that injected fluids, and fluids requiring disposal, can pose considerable hazard potentials. In summary, the study concludes that currently missing knowledge and data prevent a profound assessment of the risks and their technical controllability in Germany.

Recommendations

In light of the shortcomings of the currently available data, and of the fact that environmental risks cannot be ruled out, we recommend from a standpoint of precautionary water resources management, that above-ground and below-ground activities for unconventional gas exploitation involving fracking should not be approved for exploration or exploitation in water protection areas (classes I through III), water-extraction areas for the public drinking water supply (even if not assigned as water protection areas), mineral spa protection zones, and near mineral water reserves. These areas should be excluded for such activities. This recommendation on denial of approval should be reviewed as more data become available. In areas known to have unfavourable geological and hydrogeological conditions (groundwater potentials and known impact pathways), no exploration and exploitation of unconventional gas (via deep-drilling and hydraulic fracturing) should be allowed.

Site-specific risk assessment should be carried out with regard to any future drilling with fracking, and to drilling and use of underground disposal wells for injection of flowback. Such analyses should take account of all relevant fluids, whether introduced or encountered (fracking additives, formation water and its reaction products, and flowback), and of the relevant geological and technical impact pathways. It is recommended that use of toxicologically and ecotoxicologically hazardous fluids, and flowback disposal in disposal wells – also in the tight gas reservoirs in Germany that have already been exploited for many years – be reassessed.

Since the potential risks of exploration and exploitation of unconventional gas projects can be reliably assessed only if reliable information on the relevant geological systems (and potential impact pathways) is available, we recommend that any exploration of gas reservoirs provides investigations of the larger regional geological and hydrogeological system.

We further recommended that additional data and experience not yet published or not yet assessed (e.g., cadaster of old wells, cadaster of disposal wells) are evaluated and results are published. We argue however that without new data it will not be possible to answer the question of whether, and where, economically exploitable unconventional gas reserves are present in Germany and which technology (with or without fracking) is suited for exploration. We thus support the idea of carrying out further exploration, including exploration involving deep drilling (but without fracking), and carrying out targeted research in the above-described framework, for the purpose of answering those questions.

We recommend that further actions are taken step-by-step. Clear criteria should be established for deciding whether or not the application of fracking should be allowed at a later time. Such criteria should cover both the hazard potential of fracking additives and the availability of reliable information about the geological and technical impact pathways involved. Clear criteria should be applied for approval of any further exploration and any later production. A catalogue of criteria for approval should be developed step-by-step, applying transparent approaches involving public participation.

We recommend that research and development are intensified in areas such as the long-term integrity of wells, techniques available for forecasting the widths and lengths of fractures generated by fracking, and the development of fracking fluids with lower hazard potential. Practical application of the relevant research findings should be monitored scientifically.

With regard to EIA obligations, we recommend that fracking projects be subject to general federal EIA obligations, and that such obligations include an "opening clause" to allow participation of the German federal states. The public participation required under EIA legislation should be expanded to include a project-monitoring component, since many findings regarding projects' potential environmental impacts cannot be obtained until the projects are actually underway. Careful review of requirements under water law should be assured, via clarification of pertinent requirements, and via a) introduction of an integrated project-approval procedure to be directed by an environmental authority subordinate to the Ministry for the Environment, or b) integration of mining authorities within the environmental administration.

The following two aspects are of central importance with regard to any continuation of exploration and exploitation of unconventional gas in Germany, regardless of the procedures applied: all work processes and results should be fully transparent, and all stakeholders should exercise trust in their dealings with each other. Efforts should include the establishment of a publicly accessible cadaster listing all fracking measures carried out in the past, along with the quantities and the compositions of the fluids used.

In the following sections, we propose special recommendations for further steps towards exploitation of unconventional gas reservoirs in Germany. The focus of the recommendations is on the next phase of pilot exploration, especially, exploration in geological systems for which no information, or very little information, is yet available concerning unconventional gas reservoirs they may contain. The objectives of the recommendations include:

- identifying hydrogeological problematic areas, and possible impact pathways, at an early stage, and proposing measures for ongoing monitoring,
- reducing the hazard potential of the fracking fluid potentially used.

Special Recommendations with Regard to the Area of Geological Systems and the Aquatic Environment

The cause-and-effect relationships between deep-reaching and near-surface groundwater flow systems are of particular importance with regard to the water-related environmental impacts of unconventional gas exploitation projects. Such assessments require a detailed understanding of the hydrogeological systems involved, including:

- Conceptual hydrogeological models should be prepared that support reliable risk assessment for all potential impact pathways. The scope of such conceptual models should be large enough to support assessment of the impacts of exploration and exploitation of unconventional gas – via fracking – both for the specific sites and with regard to the large geological systems (system-oriented exploration).

- For areas in which water-related environmental impacts cannot be ruled out, numerical groundwater flow models should be prepared/refined in order to quantify the pertinent risks. This may involve preparing a regional model that can serve as a basis for local numerical models in the exploration area.

- The aforementioned numerical models have to be continually verified and calibrated on the basis of data and information obtained through monitoring (both prior and during the project). For monitoring to be effective, it must be based on an adequate understanding of the system involved. At the same time, the understanding of the system involved (conceptual or numerical model) can be improved by the monitoring data obtained. Monitoring-based project control requires meaningful indicators and an evaluation system. Ultimately, options must be available for stopping, limiting, or reversing any undesired developments. The models resulting from the aforementioned work steps provide an important basis for authorities' decisions regarding the approval of submitted projects, as well as possible ancillary provisions under water law.

- The necessary regional and local models must be provided by the mining company within the authorization procedure under mining law and water law, based on the requirements imposed by the competent mining and water authorities. A fracking project can be approved only when enough pertinent knowledge has been gained and adequate precaution has been taken to exclude any adverse impact on exploitable water resources.

Special Recommendations with Regard to the Area of Substances

Assessment of selected fracking fluids used in unconventional gas reservoirs in Germany, along with the available information on the characteristics of flowback, have revealed that injected fluids, and fluids requiring disposal, can pose considerable hazard potentials. In light of the gaps in knowledge, uncertainties and data deficits identified via research and assessment for the present study, the following recommendations for action are seen as important:

- Complete disclosure of all substances used, with regard to substance identities and quantities.

- Assessment of the toxicological and ecotoxicological hazard potentials of substances used, and provision of all physical-chemical and toxicological substance data required by the mining company. If relevant substance data are lacking, the gaps in the data must be eliminated – if necessary, via suitable laboratory tests or model calculations. In the process, the effects of relevant substance mixtures must be taken into account.

- Substitution of unsafe substances (especially, substances that are highly toxic, carcinogenic, mutagenic, or toxic for reproduction), reduction or substitution of biocides, reduction of the numbers of additives used, lowering of concentrations used.

- Determination and assessment of the characteristics of site-specific formation water, with regard to constituents of relevance to drinking water quality (salts, heavy metals, Naturally Occurring Radioactive Material – NORM, hydrocarbons).

- Determination and assessment of the characteristics of site-specific flowback, with regard to constituents of relevance to drinking water quality (salts, heavy metals, NORM, hydrocarbons), and with regard to additives used (primary substances) and their transformation products (secondary substances); determination and assessment of the proportion of fracking fluids recovered with the flowback.

- Determination of the behaviour and fate of substances in the fracking horizon, via mass balancing of the additives used.

- Modelling of substance transport, for assessment of possible risks to any exploitable groundwater, from any migrating formation water and fracking fluids.

- Technical treatment and "environmentally sound" disposal of flowback, including description of all technically feasible treatment processes and of the possibilities for reusing substances. If injecting flowback into disposal wells, conducting of a site-specific risk analysis is recommended.

- Monitoring and system-oriented examination, including installation of near-surface groundwater observation wells to determine the reference condition with regard to additives and methane; if appropriate, installation of deep groundwater observation wells to determine the characteristics of formation water and the relevant hydraulic potentials.

METHODS

Under German water law, the key requirement to be applied in assessing releases of substances into the groundwater is that releases must not adversely affect the water quality (Art. 48 (1) WHG, Federal Water Resources Management Act). An adverse effect on the quality of near-surface groundwater (i.e. of the exploitable groundwater that is integrated in natural cycles) has occurred, if water quality has worsened more than slightly.

An adverse effect on the water quality of groundwater must be assumed if relevant legal and sub-legal limit values, guide values, maximum values, and especially the "Geringfügigkeitsschwellenwerte" (de minimis thresholds, GFS) of the German Federal/State Working Group on Water (LAWA) [26] are exceeded in any exploitable groundwater. These de minimis thresholds are primarily based either on the maximum permitted concentration specified by the Ordinance on Drinking Water (Trinkwasserverordnung), or, if no maximum permitted concentration has yet been established, on toxicologically and ecotoxicologically derived threshold values. Thus, it is ensured that groundwater remains available as drinking water resource for human consumption, and it remains intact as a habitat and as part of natural cycles.

For the majority of the substances used as fracking additives, no de minimis thresholds or other water-law-based assessment values have yet been established. Therefore, hygienic guidance values for drinking water (GVDW – maximum concentration of a substance in drinking water that can be tolerated for a lifetime without suffering adverse effects on health) or health orientation values (HOV - precautionary value for substances that cannot (or can only partially) be toxicologically assessed [27]) and ecotoxicologically established Predicted No Effect Concentrations (PNEC - maximum concentration of a substance at which no effects on organisms of an aquatic ecosystem are expected [28]) were assessed for such substances, or derived using published methods, following the concept of LAWA[26].

Relevant for the assessment is the concentration at the location where the substance enters exploitable groundwater resources. In case of substances entering groundwater from the surface (pathway group 0, e.g. accidents during transport and preparation of fracking fluids), the relevant substance concentration for the assessment is the concentration at the groundwater surface (see page water). By analogy, in the case of a possible release from the fracking horizon (and related migration via pathway groups 1 through 3), the concentration at the base of the exploitable groundwater aquifer should be used in the assessment.

The relevant substance concentrations can properly assessed only site-specifically. For potential migration and exposure scenarios, suitable models are needed that consider relevant hydraulic and geochemical transport, mixing, decomposition, and reaction processes along the underground flow pathway. No such models are available at present that have the necessary spatial resolution.

As long as suitable models are lacking, we propose to assess hazard potentials on the basis of substance concentrations in (undiluted) fracking fluids and formation water. Based on the current state of knowledge, we consider it not suitable to presume a considerable reduction of their hazard potential due to dilution along the underground flow pathways, because along the flow path dilution occurs mainly by mixing with saline groundwater, which can have considerable hazard potential of its own (see below); thus, mixing with such water would not necessarily reduce the hazard potential of fracking fluids.

The pertinent hazard potentials of the fluids are assessed on the basis of the individual constituents, calculating substance-specific risk quotients of substance concentrations and assessment values (GFS, GVDW, HOV, or PNEC):

$$\mathrm{Risk\,Quotient} = \frac{\text{substance concentration in the fluid}}{\text{assessment value}}$$

When a substance has a risk quotient < 1, no hazard potential is expected, while a risk quotient ≥ 1 represents potentially a toxicological or ecotoxicological hazard (hazard potential). In the present study, a risk quotient $> 1{,}000$ is assumed to represent a high hazard potential. This value is given as an example and has not been scientifically established; it needs to be site-specifically reviewed on the basis of exposure scenarios – using numerical models for example.

AUTHORS' CONTRIBUTIONS

The authors contributed in equal parts to this publication. All authors read and approved the final manuscript.

ACKNOWLEDGEMENTS

The authors would like to thank the German Federal Environment Agency (UBA) for financing the study and the project partners [Gassner, Groth, Siederer & Coll.] and TU Darmstadt (Prof. Dr. Sass) for their collaboration.

REFERENCES

1. Bundesanstalt für Geowissenschaften und Rohstoffe (2012) Abschätzung des Erdgaspotenzials aus dichten Tongesteinen (Schiefergas) in Deutschland. Hannover.http://www.bgr.bund.de/DE/Themen/Energie/Downloads/BGR_Schiefergaspotenzial_in_Deutschland_2012.pdf?__blob=publicationFile&v=7

2. Sachverständigenrat für Umweltfragen: Fracking zur Schiefergasgewinnung (2013) Ein Beitrag zur energie- und umweltpolitischen Bewertung. Aktuelle Stellungnahme Nr. 18. Berlin.http://www.umweltrat.de/SharedDocs/Downloads/DE/04_Stellungnahmen/2012_2016/2013_05_AS_18_Fracking.pdf?__blob=publicationFile

3. Bundesanstalt für Geowissenschaften und Rohstoffe (2013) Energiestudie 2013. Reserven, Ressourcen und Verfügbarkeit von Energierohstoffen. Hannover. p 112.http://www.bgr.bund.de/DE/Themen/Energie/Downloads/Energiestudie_2013.pdf?__blob=publicationFile&v=5

4. Umweltbundesamt (2012) Umweltauswirkungen von Fracking bei der Aufsuchung und Gewinnung von Erdgas aus unkonventionellen Lagerstätten – Risikobewertung, Handlungsempfehlungen und Evaluierung bestehender rechtlicher Regelungen und Verwaltungsstrukturen. -Gutachten im Auftrag des Umweltbundesamtes. Berlin.http://www.umweltbundesamt.de/uba-info-medien/4346.html

5. Ministerium für Klimaschutz, Umwelt, Landwirtschaft, Natur- und Verbraucherschutz des Landes NRW (2012) Gutachten mit Risikostudie zur Exploration von Erdgas aus unkonventionellen Lagerstätten in Nordrhein-Westfalen und deren Auswirkungen auf den Naturhaushalt, insbesondere die öffentliche Trinkwassergewinnung.Düsseldorf.http://www.umwelt.nrw.de/umwelt/wasser/trinkwasser/erdgas_fracking

6. IWW Rheinisch-Westfälisches Institut für Wasser Beratungs- und Entwicklungsgesellschaft mbH (2013) Wasserwirtschaftliche Risiken bei Aufsuchung und Gewinnung von Erdgas aus unkonventionellen Lagerstätten im Einzugsgebiet der Ruhr. Gutachten des IWW im Auftrag der Arbeitsgemeinschaft der Wasserwerke an der Ruhr e.V. und des Ruhrverbandes. Mülheim. http://www.awwr.de/fileadmin/download/download_2013/studie_fracking_einzugsgebiet_ruhr.pdf, http://www.ruhrverband.de/wissen/forschung-entwicklung/fracking/

7. U.S. Environmental Protection Agency (2004) Evaluation of impacts to underground sources of drinking water by hydraulic fracturing of coalbed methane reservoirs.http://water.epa.gov/type/groundwater/uic/class2/hydraulicfracturing/wells_coalbedmethanestudy.cfm

8. U.S. Environmental Protection Agency (2011) Plan to Study the Potential Impacts of Hydraulic Fracturing on Drinking Water Resources. Washington.http://water.epa.gov/type/groundwater/uic/class2/hydraulicfracturing/upload/hf_study_plan_110211_final_508.pdf

9. Tyndall Centre (2011) Shale gas: a provisional assessment of climate change and environmental impacts. Manchester.http://www.tyndall.ac.uk/shalegasreport

10. Waxman HA, Markey EJ, Degette D (2011) Chemicals used in hydraulic fracturing. In: U.S. House of Representatives Committee on Energy and Commerce Minority Staff.Washington. http://democrats.energycommerce.house.gov/sites/default/files/documents/Hydraulic-Fracturing-Chemicals-2011-4-18.pdf

11. New York State Department of Environmental Conservation (2011) Revised Draft Supplemental Generic Environmental Impact Statement. Chapter 5: Natural gas development activities & high-volume hydraulic fracturing. New York.http://www.dec.ny.gov/docs/materials_minerals_pdf/rdsgeisch50911.pdf

12. Ewen C, Borchardt D, Richter S, Hammerbacher R (2012) Risikostudie Fracking – Übersichtsfassung der Studie "Sicherheit und Umweltverträglichkeit der Fracking-Technologie für die Erdgasgewinnung aus unkonventionellen Quellen" erstellt im Zusammenhang mit dem InfoDialog Fracking. Darmstadt.http:// dialog-erdgasundfrac.de/sites/dialog-erdgasundfrac.de/files/Ex_ risikostudiefracking_120518_webprint.pdf

13. Hessischer Landtag (2013) 60. Sitzung des Ausschusses für Umwelt, Energie, Landwirtschaft und Verbraucherschutz. Wiesbaden.http://www.hessischer-landtag.de/icc/med/bb7/bb700690-9433-e31a-628b-31402184e373,11111111-1111-1111-1111-111111111111. pdf

14. Landesamt für Bergbau, Energie und Geologie Niedersachsen Mindestanforderungen an Betriebspläne, Prüfkriterien und Genehmigungsablauf für hydraulische Bohrlochbehandlungen in Erdöl- und Erdgaslagerstätten in Niedersachsen. Clausthal-Zellerfeld.

15. Rundverfügung vom 31.10.2012.http://www.lbeg. niedersachsen.de/download/72198/Mindestanforderungen_ an_Betriebsplaene_Pruefkriterien_und_Genehmigungsablauf_ fuer_hydraulische_Bohrlochbehandlungen_in_Erdoel-_und_ Erdgaslagerstaetten_in_Niedersachsen.pdf

16. Hammerbacher Beratung & Projekte Statusbericht von ExxonMobil zur Umsetzung der Risikostudie Fracking. Osnabrück.Protokoll vom 6. November 2012, Osnabrück.http://www.erdgassuche-in-deutschland.de/dialog/info_dialog_fracking_status.html

17. Ministerium für Klimaschutz, Umwelt, Landwirtschaft, Natur- und Verbraucherschutz des Landes Nordrhein-Westfalen Pressemitteilung vom 07.09.2012 - Umweltministerium und Wirtschaftsministerium legen Risikogutachten zu Fracking vor. http://www.umwelt.nrw.de/ministerium/service_kontakt/archiv/ presse2012/presse120907_a.php

18. Remmel J (2012) Erdgas aus unkonventionellen Lagerstätten. gwf Wasser Abwasser 11:1121

19. Niedersächsisches Ministerium für Wirtschaft, Arbeit und Verkehr Gemeinsame Presseinformation von Minister Wenzel und Lies vom 17.03.2014 - Ja zur Erdgasförderung! Nein zu umwelttoxischen

Substanzen unter Tage!http://www.mw.niedersachsen.de/portal/
live.php?navigation_id=5459&article_id=123032&_psmand=18

20. Bundesministerium für Umwelt, Naturschutz und Reaktorsicherheit Vorschlag zur Änderung von UVP-V und Wasserhaushaltsgesetz.http://www.bmu.de/themen/wasser-abfall-boden/binnengewaesser/gesetzesaenderung-zu-fracking

21. Deutscher Verein des Gas- und Wasserfaches e.V Stellungnahme vom 21. März 2013 zum Entwurf eines Gesetzes zur Änderung des Wasserhaushaltsgesetzes vom 7. März 2013 und Entwurf einer Verordnung zur Änderung der Verordnung über die Umweltverträglichkeitsprüfung bergbaulicher Vorhaben vom 11. März 2013 in Bezug auf die Umweltverträglichkeitsprüfung bei Bohrungen mit Einsatz der Fracking-Technologie.http://www.dvgw.de/wasser/ressourcenmanagement/gewaesserschutz/fracking/

22. BGR, GFZ & UFZ (2013) Abschlusserklärung zur Konferenz "Umweltverträgliches Fracking?". Hannover.am 24./25. Juni 2013 (Hannover-Erklärung).http://www.bgr.bund.de/DE/Gemeinsames/Nachrichten/Veranstaltungen/2013/GZH-Veranst/Fracking/Downloads/Hannover-Erklaerung-Finalfassung.pdf

23. Gelsenwasser AG, Arbeitsgemeinschaft der Wasserwerke an der Ruhr e.V., Deutscher Brauer–Bund e.V., Verband Deutscher Mineralbrunnen e.V. & Wirtschaftsvereinigung Alkoholfreie Getränke e.V (2013) Gelsenkirchener Erklärung: Wasserversorger, Bierbrauer, Mineral– und Heilbrunnenbetriebe sowie Erfrischungsgetränkehersteller fordern Schutz vor Fracking. Gelsenkirchen.

24. (24.10.2013)http://www.gelsenwasser.de/fileadmin/download/unternehmen/presse/gelsenkirchener_erklaerung.pdf

25. ExxonMobil Central Europe Holding GmbH Frack-Flüssigkeiten. http://www.erdgassuche-in-deutschland.de/erkundung_foerderung/frac_fluessigkeiten/index.html

26. Bezirksregierung Arnsberg (2011) Gewinnung von Erdgas aus unkonventionellen Lagerstätten – Erkundungsmaßnahmen der CONOCO Mineralöl GmbH in den Jahren 1994 – 1997. Arnsberg.61.01.25-2010-9.

27. Rosenwinkel KH, Weichgrebe D, Olsson O (2012) Gutachten Stand der Technik und fortschrittliche Ansätze in der Entsorgung

des Flowback des Instituts für Siedlungswasserwirtschaft und Abfall (ISAH) der Leibniz-Universität Hannover zum Informations- und Dialogprozess über die Sicherheit und Umweltverträglichkeit der Fracking-Technologie für die Erdgasgewinnung. Hannover. http://dialog-erdgasundfrac.de/sites/dialog-erdgasundfrac.de/files/Gutachten%20zur%20Abwasserentsorgung%20und%20 Stoffstrombilanz%20ISAH%20Mai%202012.pdf

28. LAWA – Bund/Länder-Arbeitsgemeinschaft Wasser Ableitung von Geringfügigkeitsschwellen für das Grundwasser. Düsseldorf: 2004. http://www.lawa.de/documents/GFS-Bericht-DE_a8c.pdf

29. Umweltbundesamt (2003) Bewertung der Anwesenheit teil- oder nicht bewertbarer Stoffe im Trinkwasser aus gesundheitlicher Sicht. Empfehlung des Umweltbundesamtes nach Anhörung der Trinkwasserkommission des Bundesministeriums für Gesundheit. Bundesgesundheitsbl Gesundheitsforsch Gesundheitsschutz 46:249-251

30. European Commission (2003) Technical Guidance Document in support of Commission Directive 93/67/EEC on Risk Assessment for new notified substances, Commission Regulation (EC) No 1488/94 on Risk Assessment for existing substances and Directive 98/9/EC of the European Parliament and of the Council concerning the placing of biocidal products on the market, Part II. Ispra. http://ihcp.jrc.ec.europa.eu/our_activities/public-health/risk_assessment_of_Biocides/doc/tgd.

Ecological Characterization of Soil-inhabiting and Hypolithic Soil Crusts within the Knersvlakte, South Africa

Bettina Weber[1, 2], Dirk CJ Wessels[3], Kirstin Deutschewitz[4], Stephanie Dojani[1], Hans Reichenberger[1], and Burkhard Büdel[1]

Bettina Weber[1, 2], Dirk CJ Wessels[3], Kirstin Deutschewitz[4], Stephanie Dojani[1], Hans Reichenberger[1], and Burkhard Büdel[1]

[1]Department of Biology, University of Kaiserslautern, Plant Ecology and Systematics, Kaiserslautern 67653, Germany

[2]Now at: Department of Multiphase Chemistry, Max Planck Institute for Chemistry, Hahn-Meitner-Weg 1, Mainz 55128, Germany

[3]Department of Biodiversity, School of Molecular and Life Sciences, University of Limpopo, Turfloop Campus, Private, Sovenga 0727, South Africa

[4]URS Deutschland GmbH, Europaallee 3-5, Kaiserslautern 67657, Germany

ABSTRACT

Introduction

Within the Knersvlakte, cyanobacteria occur hypolithically underneath translucent quartz stones in areas with quartz pavement and, outside pavement areas, they are soil-inhabiting within the uppermost millimeters of the soil. Both habitats were characterized in terms of biomass and growth patterns of cyanobacteria. Long-term microclimatic conditions were determined.

Methods

Biomass of organisms within both habitats was determined by means of chlorophyll analyses. A transect approach was used to determine the frequency of hypolithic growth depending on the size, weight, and embedding depth of the quartz pebbles. Organisms were identified by means of microscopic analyses of the samples. Microclimatic conditions within both habitats, i.e., temperature, light intensity, air humidity, and soil moisture, were recorded bi-hourly from September 23, 2004 through September 7, 2006.

Results

The biomass of hypolithic and soil-inhabiting crusts was almost identical, 88 vs. 86 mg Chl_a/m^2 and 136 vs. 134 mg Chl_{a+b}/m^2. Within the quartz fields, 46.8% of the surface area was covered by quartz stones with 69% of translucent quartz stones colonized by hypolithic cyanobacteria and algae. Colonized quartz stones were significantly thicker, heavier, and more deeply embedded in the soil than uncolonized ones. Whereas the annual mean temperature on top of quartz stones was nearly identical to that underneath thin and thick quartz stones, daily temperature amplitudes were largest on the stone surface (up to 48.1 K), compared to the hypolithic habitats (up to 39.4 K). Light intensity in the hypolithic habitat was between 15 and 30% of the ambient light intensity during daytime. Water condensation in the absence of rain occurred during 50% of the nights on the quartz stone surface, but only during 34% of the nights on the soil surface during winter months within 1 year. Soil moisture beneath quartz layers was greater and less variable than beneath soil-inhabiting crusts.

Conclusions

In spite of the large differences in the microclimatic conditions, both habitats seem to be similarly well suited for cyanobacterial growth, resulting in equal biomass values but some differences in taxonomic composition.

INTRODUCTION

Biological soil crusts (BSC) are well known to occur in hot and cold deserts and semi-deserts throughout the world. They comprise cyanobacteria, algae, lichens, and bryophytes as well as bacteria and fungi in variable proportions growing within the uppermost millimeters of the soil (Belnap et al. 2003). Based on the successionally most advanced organism group present, they are coarsely defined as cyanobacteria-, lichen-, and bryophyte-dominated soil crusts (Büdel et al. 2009). In addition to these more general types of soil-inhabiting crusts, Büdel et al. (2009) define two additional types of BSC for southern Africa: the Namib lichen fields, which are restricted to this fog desert environment, and the hypolithic crusts. The latter are mostly dominated by cyanobacteria, often accompanied by green algae, diatoms, fungi, and bacteria (Cameron 1960; Cameron and Blank1967; Friedmann et al. 1967; Friedmann and Galun 1974; Broady 1981; Rummrich et al. 1989; Büdel and Wessels 1991; Schlesinger et al. 2003; Galun and Garty 2003; Ullmann and Büdel 2003; Cockell and Stokes 2004; Bhatnagar and Bhatnagar 2005). Lichens (Büdel and Schultz 2003), bryophytes (Müller 2009; Cowan et al. 2010), and even red-colored chloroflexi bacteria (Lacap et al.2011) have also been reported to form the dominant organism group on the bottoms and sides of stones which are tightly embedded into the substrate. Hypolithic growth has been recorded on the soil-embedded sides and underneath quartz (e.g., Vogel 1955; Pointing et al. 2007; Schlesinger et al. 2003), limestone (Friedmann et al. 1967), dolomite (Cockell and Stokes 2006), marble (Cowan et al. 2010), flint stone (Friedmann et al. 1967; Berner and Evenari 1978), prehnite and agate (Tracy et al. 2010), and even calcareous material of organic origin such as bivalve shells and recently dead snails (Friedmann et al. 1967), all of which are somewhat translucent, allowing enough light to reach the niche-inhabiting organisms.

The hypolithic habitat is known to be characterized by different ecological conditions compared to those occurring on the soil surface, i.e., it experiences lower intensities of overall light and UV radiation (Vogel 1955; Berner and Evenari 1978; Schlesinger et al. 2003; Cowan et al. 2010), buffered thermal conditions (Broady 1981; Warren-Rhodes et al. 2006), and enhanced moisture availability compared to the surrounding soil (Warren-Rhodes et al. 2006). In contrast to that, Schlesinger et al. (2003) also observed lower light and improved water conditions, but higher temperature amplitudes and absolute temperature values and described that as a modest greenhouse effect for the hypolithic community of the Mojave Desert.

At our study site in the Knersvlakte, we observed widely developed hypolithic growth of cyanobacterial organisms (Figure 1A, C; Weber et al. 2010), but also well-developed cyanobacteria-dominated soil-inhabiting BSC in regions without quartz stone coverage (Figure 1B, D). In order to analyze the utilization of both ecological niches by cyanobacteria and algae, we identified cyanobacterial genera of major relevance and determined the biomass of photosynthetically active organisms by means of chlorophyll analyses. Utilization of the hypolithic habitat was characterized through quantification and measurement of hypolithic growth parameters. By means of long-term meso- and microclimate measurements, we analyzed and compared the conditions within both habitats. The effects of these microclimatic differences on long-term carbon fixation patterns are discussed herein.

Figure 1: The Knersvlakte and its habitats for cryptogamic organisms. (A) Overview of the Knersvlakte with extended quartz fields and the characteristic flora, which comprises highly diverse succulent plants and dwarf shrubs. (B) Within the Knersvlakte, large tracts of quartz fields with hypolithic growth occur next to soil surfaces, facilitating epedaphic growth of biological soil crusts. (C) Quartz pebbles of the Knersvlakte with hypolithic cyanobacterial growth on the formerly embedded part of the quartz stone. (D) Cyanobacteria-dominated biological soil crusts occurring in quartz-free regions within the Knersvlakte.

METHODS

Study Site

This study was conducted on the farm Goedehoop, next to the BIOTA South Observatory Goedehoop (No. 26) in the Van Rhynsdorp District, Western Cape Province, South Africa. The farm lies in a semiarid area known as the Knersvlakte, characterized by an extensive gently rolling gravely plain. The vegetation and other characteristics of the Knersvlakte have been reported by Cowling et al. (1999). Large tracts of the Knersvlakte are covered by a layer of quartz stones (Figure 1A), within which a distinct quartz field flora occurs (Schmiedel 2002). The importance of these quartz fields, which vary in size, was stressed by Schmiedel (2002), who pointed out that they represent a biodiversity hotspot with 150 taxa (species and subspecies), 121 (80.7%) of them being endemic and occurring obligatorily in quartz fields. The quartz fields are interspersed with locally confined distinct dwarf shrubbery. The Knersvlakte experiences cool winters and hot summers, with daily air temperatures that range from maximum values of 47.9°C in February to minimum values of 1.1°C in June (BIOTA meteorological station readings from 2001 to 2009; Jürgens et al. 2010). Although rainfall occurs throughout the year, the bulk of the annual rainfall occurs during the cool winter months (July–August), with smaller peaks occurring during April (autumn) and October (early summer). The mean total annual rainfall for the period of measurement at the BIOTA meteorological station at Ratelgat (31.283°S, 18.603°E, height 240 m a.s.l.) was 125.7 mm, ranging from 69.2 mm (in 2003) to 175.4 mm (in 2008). Detailed mesoclimatic conditions of the study site, which were measured by us, are described in the Results section.

Microscopical Determination

For determination of the most common genera of cyanobacteria, samples of soil-inhabiting crust (belonging to the following crust types: 1: light cyanobacterial crust, 2: well-established cyanobacterial crust, and 6: hypolithic crust; Büdel et al. 2009) were randomly collected in Petri dishes within the study area. Light microscopy was used to determine the soil crust material, based on Geitler (1932) and Komárek and Anagnostidis (1998, 2005).

Biomass

For chlorophyll determination of the soil-inhabiting crust, one Petri dish sample (covering 57 cm^2) was collected from each of five independent randomly chosen soil crust sampling points. For determination of the hypolithic community, 11 Petri dish samples were randomly collected within the quartz field areas. Quartz pebbles were small enough to allow for this sampling procedure. The lower part of a Petri dish was lined with cellulose paper, pressed upside down into the substrate and then carefully removed from the underlying soil with a trowel. Surplus soil was removed, and Petri dishes were then closed with the upper lid and sealed. Additionally, the hypolithic crust of five randomly chosen squares within the quartz fields, each covering an area of 19.7 × 19.7

cm, was sampled to a depth of about 3 cm and transported in double-lined polyethylene bags. In both types of samples, the complete soil surface together with the quartz gravel was collected in a dry state for later analysis in the laboratory. Chlorophyll (Chl) was extracted two times in 100% DMSO (dimethylsulfoxide) for 90 min at 65°C according to the method described by Ronen and Galun (1984). $Mg_2(OH)_2CO_3$ was added to avoid acidification and concomitant pheophytinization of chlorophyll. After spectrophotometry, the Chl_a content was calculated according to Arnon et al. (1974). The Chl_{a+b} content was calculated according to Lange, Bilger, and Pfanz (pers. comm.) as follows:

$$Chl_{a+b}[\mu g] = [20.2 * (E_{648} - E_{700}) + 8.02 * (E_{665} - E_{700})] * a \tag{1}$$

And

$$Chl_a[\mu g] = [12.19 * (E_{665} - E_{700})] * a \tag{2}$$

Where

$Chl_{a+b}[\mu g]$, $Chl_a[\mu g]$ is the chlorophyll content of the sample in micrograms, E_{648}, E_{665}, E_{700} are the absorptions at the given wavelengths, and a is the amount of DMSO used in milliliters.

The chlorophyll content per surface area was calculated by dividing the Chl value by the surface area of the sample. Chl contents of both habitats proved to be normally distributed, as checked by the Kolgomorov-Smirnov and Shapiro-Wilk tests. A T-test was used to analyze differences in Chl contents between the two habitats.

Quartz Stone Coverage and Size-Dependent Colonization

North of the Goedehoop observatory (31.2750°S, 18.6003°W), the quartz stone coverage and size-dependent colonization were determined along a 300 m transect with 30 sampling points. At each sampling point, all pebbles along a 1 m line were collected, separated into stones with and without hypolithic growth, and packaged for later analysis in the laboratory. At each point a digital image of 1 m² soil surface was taken for later analysis of quartz stone density.

All colonized and uncolonized quartz stones per sampling point were counted, weighed, and their thickness and depth of embedding measured with a digital caliper. Irregularly shaped stones were measured at their thickest and widest points. Embedded depth was measured at the deepest point, thus determining the deepest possible point of colonization by hypoliths. A Kruskal-Wallis one-way analysis of variance on ranks (SigmaStat Ver. 3.1) was used to determine the significance of differences between the thickness, mass, and embedded depth of colonized and uncolonized quartz stones.

Quartz stone density analysis was done by digital analysis of the 1 m^2 images using Adobe Photoshop software. In a first step, all vascular plants with clear bounds (e.g., *Agyroderma* sp.) were masked and saved as an extra image. The remaining image was displayed with maximum contrast as a black and white image, the soil being shown in black and the pebbles in white. The coverage of each color was analyzed with ArcView 3.2 (ESRI Inc., Redlands, WA, USA) software and the Spatial Analyst application. All images were converted into grids, in which black pixels were assigned the value 0 and white ones the value 1. The number of pixels with each value was extracted from the attribute table and converted into percentages. The images with the vascular plants were processed in the same manner and their percentage subtracted from the total pixel number per image, yielding the percentages of quartz pebbles and bare soil. In order to validate this digital method, three 25 × 25 cm areas of one sampling image were printed in color, and on each print, two 5 × 5 cm areas were randomly covered with transparent scale paper. On an illuminated screen, the edges of the quartz pebbles were traced with a pen onto the 5 × 5 cm scale papers, which were then enlarged and the quartz pebble pixels were counted. A comparison of the two methods revealed a maximum difference of 3.8%.

Generally, only images that were taken before 4 p.m. could be included in the analysis, as shadows in the image distorted the results in those taken later. Images with shrubs were also excluded from the analysis since it was not possible to mask them out in a detailed manner. Consequently, ten 1 m^2 plots could be used for the quartz pebble density analysis.

Meso- and Microclimate

The meso- and microclimate weather station was installed at 31.275°S, 18.598°E, (height 196.5 m a.s.l.), adjacent to the northern edge of the BIOTA observatory Goedehoop (Observatory No. 26), between an area whose soil surface is totally covered by quartz stones (quartz field) and an area whose soil surface is either bare or covered by a biological soil crust. A number of *Argyroderma pearsonii* (N.E. Brown) Schwantes (Aizoaceae) plants grow in the studied quartz field, with numerous other plant species, all dwarf in habit, occurring in soil without a quartz stone cover adjacent to the quartz field. At the weather station the following mesoclimatic parameters were assessed: air temperature and air humidity at a height of 100 cm, protected from direct insolation by an inhouse-built radiation shield; precipitation (Davis Rain Collector II, Davis Instruments, Hayward, CA, USA); and ambient photosynthetically active radiation (PAR; LI-190, Li-COR, Lincoln, NE, USA).

Temperature

Microclimatic thermal conditions within the quartz field were determined through the placement of one small thermistor (NTC, close tolerance R/T curve matched thermistors) per location set out below (Table 1). Temperature changes within a quartz stone were determined by placing a thermistor snugly within a 5 mm deep hole drilled into the stone. The opening of the hole was sealed off with Pratley® Quickset Putty.

Table 1: Location of small thermistors determining the thermal conditions on the soil surface and within the quartz field

Probe location (all stones with hypolithic communities)	Stone size [length × width × thickness (mm)]	Abbreviation
Below a thin quartz stone	32.57 × 20.16 × 12.01	Thin stone
Below a thicker quartz stone	33.98 × 44.45 × 21.69	Thick stone
Within a quartz stone	34.51 × 22.50 × 20.50	Inside stone
Upper surface of a quartz stone		Stone surface
Surface of biological soil crust		Crust surface

Weber et al.

Weber et al. Ecological Processes 2013 2:8, doi:10.1186/2192-1709-2-8

Light

A light sensor (G1116, Hamamatsu, Hamamatsu City, Japan) was installed under each of two quartz stones (length × width × height: stone 1: 35 × 28 × 10 mm; stone 2: 26 × 20 × 9 mm). Both stones were colonized by hypolithic cyanobacteria, thus representative of the prevailing hypolithic habitat. The organisms were then removed from the lower portion of the stones, and sensors were glued onto the undersides of these stones to measure the amount of light reaching the hypolithically growing organisms. Unfortunately, one sensor degraded after 6 months of measurements (due to muddy water leaking into the sensor on October 21, 2005). The second sensor remained intact and stable for the whole measuring time.

Condensation

The potential condensation time on the surface of quartz stones and on the soil surface was estimated for the time span from May 13 to October 19, 2005. Calculations were based on a rule of thumb for calculating the saturating vapor pressure, the so-called Magnus equation (eq. 3). The vapor pressure is defined as the product of air humidity and saturating vapor pressure (eq. 4). The dew point temperature can thus be calculated from the relative air humidity and the present temperature (eq. 5). Potential condensation conditions occurred if the quartz pebble surface temperature or soil crust surface temperature fell below this dew point temperature.

$$T) = 6.1078 * 10((a * T)/(b + T)) \tag{3}$$

$$VP\,(r, T) = r/100 * SVP\,(T) \tag{4}$$

$$TD(r, T) = b * \frac{v}{(a-v)} \text{ where } v(r, T)$$
$$= \log10(VP(r.T)/6.1078) \tag{5}$$

Where

r = relative air humidity

T = temperature in °C

TD = dew point temperature in °C

VP = vapor pressure in hPa

SVP = saturating vapor pressure in hPa

Parameters:

a = 7.5, b = 237.3 at $T \geq 0$

a = 7.6, b = 240.7 at $T < 0$ over water (dew point)

a = 9.5, b = 265.5 at $T < 0$ over ice (frost point)

The latter two sets of parameters were not used, since temperatures always remained above 0°C.

In a first approach, the total duration of potential condensation conditions within the hypolithic and the soil-inhabiting crust habitats was calculated. In a second approach, only condensation conditions at least 1 h after the last precipitation event were considered. These data were then used to calculate the number of nights (from dawn to dusk) when condensation conditions occurred.

Soil Moisture

Soil moisture was measured with two humidity sensors (ECH$_2$O EC-10, Decagon Devices, Pullman, WA, USA) at a depth of 30 mm within the soil. One sensor was installed below a soil crust surface, and the second one was placed below a quartz layer with hypolithically growing cyanobacteria.

Signals from all sensors were recorded every half hour by an in-house-built data logger from September 23, 2004 through September 7, 2006.

RESULTS

Microscopical Determination

Within the Knersvlakte, *Chroococcidiopsis, Leptolyngbya, Lyngbya, Microcoleus, Nostoc, Oscillatoria, Phormidium, Pseudanabaena, Pseudophormidium, Schizothrix, Scytonema, Tolypothrix,* and *Trichocoleus* were found to be the dominant cyanobacterial genera. The genus *Chroococcidiopsis* was observed to occur exclusively in the hypolithic environment, whereas *Microcoleus* and *Nostoc* were mainly found in soil-inhabiting crust samples.

Biomass

Within the soil-inhabiting crust and hypolithic habitat, nearly identical biomass values were present with Chl_a contents of 86.1 mg Chl_a/m^2 (SD = 37.8; $n = 5$) and 87.7 mg Chl_a/m^2 (SD = 38.8; $n = 16$), respectively. Chl_{a+b} contents were about 2/3 larger at 133.7 mg Chl_{a+b}/m^2 (SD = 59.9; $n = 5$) in soil crusts and 136.2 mg Chl_{a+b}/m^2 (SD = 52.4; $n = 16$) in hypolithic crusts. There were no significant differences in the Chl values between the two habitats.

Quartz Stone Coverage and Size-dependent Colonization

Within the quartz fields of Goedehoop, on average 46.8% (SD: 2.6, $n = 10$) of the soil surface was covered by quartz stones. Hypolithic communities were present beneath 69% of the sampled quartz stones ($n = 682$).

Colonized quartz stones were significantly thicker than uncolonized ones ($P < 0.001$) with a mean thickness of 9.5 mm (median = 8.8 mm, SD = 4.2, $n = 363$) compared to 5.7 mm (median = 5.3 mm, SD = 2.5, $n = 319$) for uncolonized pebbles. The largest number of colonized stones had a thickness of 9 mm (Figure 2A), whereas the largest number of uncolonized pebbles were 6 mm thick. The colonized quartz stones were up to 29.8 mm thick; uncolonized stones only reached a maximum thickness of 18.3 mm. The mean thickness of all quartz stones from the sampled quartz field was 7.8 mm and ranged in thickness from 1.7 to 29.8 mm.

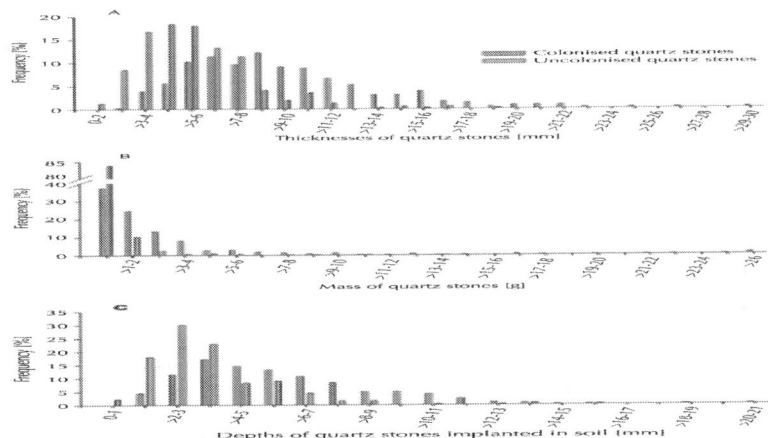

Figure 2: Frequency histograms of the thickness (A), mass (B) and depth embedded in the soil (C) of colonized and uncolonized quartz stones from a quartz field in the Knersvlakte, Western Cape, South Africa. The mean values of colonized and uncolonized quartz stones were significantly different ($P <$ 0.001) for all three parameters.

The mass of colonized quartz stones was significantly higher than that of uncolonized pebbles ($P < 0.001$), although the largest number of quartz stones with hypolithic growth had a mass ranging from more than 0.08 to 1.0 g (Figure 2B). Hypolithic communities occurred underneath quartz stones with a mass of up to 138.6 g, whereas the maximum mass of uncolonized stones was 13.4 g. The mean mass of colonized quartz stones from the sampled quartz field was 3.4 g (median = 1.5 g, SD = 8.8, $n = 363$), that of uncolonized quartz pebbles was 0.7 g (median = 0.4 g, SD = 1.3, $n = 319$).

Colonized quartz stones were embedded significantly deeper in the soil than uncolonized quartz stones ($P < 0.001$) with a mean depth of 5.8 mm (median = 5.2 mm, SD = 3.0, $n = 358$) for colonized and 3.5 mm (median = 3.0 mm, SD = 2.1, $n = 268$) for uncolonized quartz stones. The colonized quartz stones reached up to a maximum depth of 20.1 mm within the soil, whereas uncolonized pebbles were embedded up to 15.9 mm (Figure 2C). The largest numbers of both colonized and uncolonized quartz stones were embedded 4 mm into the soil, ranging from a minimum of 0.4 mm to a maximum of 20.1 mm.

Mesoclimate

The mesoclimatic characteristics at the Knersvlakte site from September 1, 2005 to August 31, 2006 revealed a typical southern hemispherical climate with the highest daily mean temperature (32.3°C) being reached on February 1 and the lowest (9.2°C) on July 22 (Figure 3). The maximum air temperature recorded during that year was 44.8°C (February 1); the absolute minimum of −1.2°C was measured on June 6. Highest light intensities were measured during the summer months from November through February (Figure 3) with maximum light intensities of 2,251 μmol PAR m⁻² s⁻¹being recorded on February 7; lowest values were recorded from June through July with the lowest daily maximum (433 μmol PAR m⁻² s⁻¹) occurring on August 24. The total amount of precipitation was 151.8 mm with the highest monthly precipitation amounts in May (65.2 mm), June (20.2 mm), July (12.8 mm), and August (21.4 mm), and the lowest amounts during summer in November (2 mm) and December (0.2 mm; Figure 3). There were no months without any rain during that year.

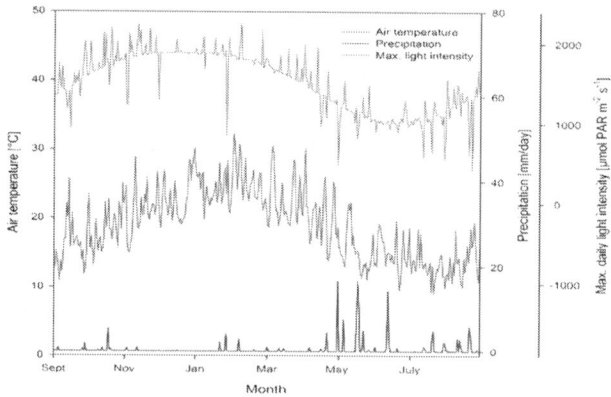

Figure 3: Mesoclimatic conditions in the Knersvlakte, Western Cape, South Africa. Daily light intensity maxima, daily mean air temperatures, and daily amounts of precipitation, from September 1, 2005 to August 31, 2006.

Microclimate

Microclimate sensors were installed in various locations to explore the microclimatic conditions within the soil-inhabiting crust and hypolithic habitat.

Temperature

Temperature conditions experienced by soil-inhabiting and hypolithically growing BSC were analyzed over the course of 1 year from September 1, 2005 to August 31, 2006. Temperatures on the surface of quartz stones (Figure 4A) and at the soil surface (Figure 4B) were highest during the day and lowest at night during most summer days. Temperature conditions within a quartz stone, under thin and thick quartz stones, and at a depth of 30 mm underneath the quartz field were increasingly moderated by the insulating effect of stone and soil. Air temperatures were almost always lower than soil/quartz temperatures, except for a few short events, as in the late evening of January 3, 2006 between 10:30 and 11:00 p.m. (quartz surface temperature higher than air temperature; Figure 4A).

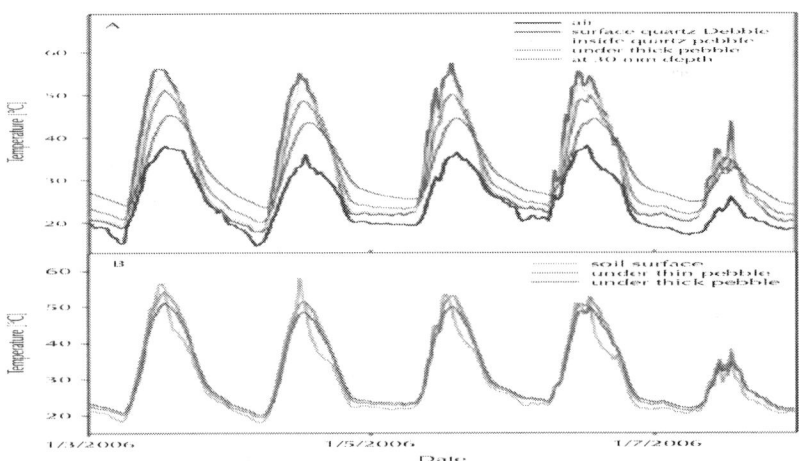

Figure 4: Summer temperatures measured at different locations during a characteristic time span, January 3–7, 2006. (A) Temperatures measured at a height of 1.80 m (ambient air temperature), at the quartz pebble surface,

inside a quartz pebble, under a thick quartz pebble, and at a depth of 30 mm underneath a quartz field. (B) Temperatures measured at the surface of epigaeic soil crusts and under a thin and a thick quartz pebble. For dimensions of quartz pebbles, see Table 1.

During winter, the same insulating effect of stone and soil was present (Figure 5). The differences between air and soil/quartz pebble temperatures, however, were far less extreme. During early morning and late evening hours, air temperatures were frequently higher than soil/quartz pebble temperatures and sometimes they were even higher during the day and night (e.g., June 24–26, 2006, Figure 5). Over the course of the whole year, highest daily maxima and lowest daily minima were reached at the soil surface and the quartz stone surface (Table 2). Thus, the highest daily amplitudes of 48.1 and 47.5 K were also reached at the quartz stone surface and soil surface, respectively, whereas under the thin and thick quartz stones, amplitudes of only 39.4 and 36.0 K were recorded, respectively (Table 2).

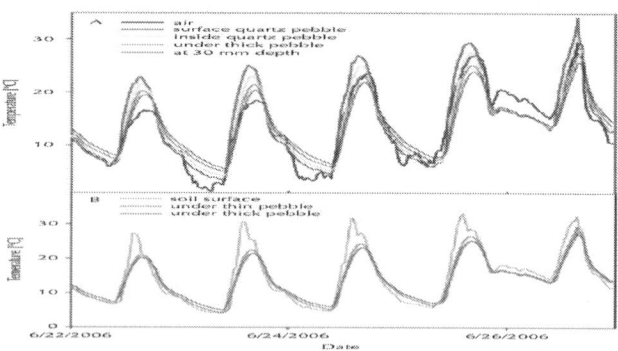

Figure 5: Winter temperatures measured at different locations during a characteristic time span, June 22–26, 2006. (A) Temperatures measured at a height of 1.80 m (ambient air temperature), at the quartz pebble surface, inside a quartz pebble, under a thick quartz pebble, and at a depth of 30 mm underneath a quartz field. (B) Temperatures measured at the surface of epigaeic soil crusts and under a thin and a thick quartz pebble. For dimensions of quartz pebbles, see Table 1.

Table 2: Daily temperature maxima, minima, and amplitudes on the quartz stone surface, underneath a thin and a thick quartz stone and at the soil crust surface

		Surface of quartz stone	Under thin quartz stone	Under thick quartz stone	
Daily maximum	Max. value [°C]	63.5	59.7	56.8	65.6
	Date	12/30/05	12/30/05	12/30/05	02/06/06
Daily maximum	Min. value [°C]	15.6	15.0	14.8	14.4
	Date	07/31/06	07/31/06	07/31/06	07/31/06
Daily minimum	Max. value [°C]	24.6	25.7	26.0	24.7
	Date	02/06/06	02/02/06	02/02/06	02/06/06
Daily minimum	Min. value [°C]	3.4	4.2	5.0	4.1
	Date	06/23/06	06/23/06	06/23/06	06/23/06
Daily amplitude	Max. value [K]	48.1	39.4	36.0	47.5
	Date	10/14/05	12/30/05	12/30/05	10/31/05
Daily amplitude	Min. value [K]	5.3	4.6	4.0	3.7
	Date	07/31/06	07/31/06	07/31/06	08/24/06

Weber et al.

Weber et al. Ecological Processes 2013 2:8, doi:10.1186/2192-1709-2-8

Temperatures on the bottom of the thin and thick quartz stones were up to 4.0 and 4.2 K higher than at the quartz stone surface (on November 6, 2005, 4:30 p.m. and September 20, 2005, 6:30 p.m., respectively) and up to 15.4 and 15.7 K lower (on October 2, 2005, 10:00 a.m. and October 14, 2005, 11:00 a.m., respectively). Over the course of the year, temperatures under the thin and thick stones were higher than those at the quartz surface on 220.8 and 224.6 days, equal for 7.8 and 3.4 days, and lower for 136.2 and 136.8 days, respectively. Over the whole year, the mean temperature values under the thin and thick quartz stones were 0.3 and 0.5 K lower, respectively. During the summer months (December through February) mean temperatures under the thin and thick stones were 0.2 and 0.6 K lower than at the quartz surface, whereas during winter months, temperature differences were 0.5 and 0.4 K lower temperatures, respectively.

Light

Ambient light intensities reached values up to 2,273 µmol PAR m^{-2} s^{-1} during summer, whereas under the quartz pebbles the maximum light intensity was only 586 µmol PAR m^{-2} s^{-1}. The course of light intensity during daylight closely followed the ambient light (Figure 6), reaching 15–30% of the ambient light intensity with higher percentages in the afternoon hours.

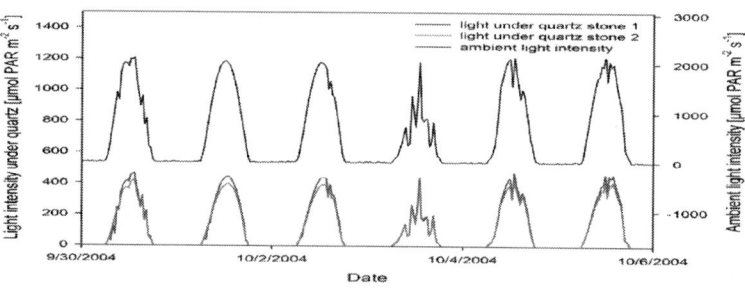

Figure 6: Light intensity measured underneath two translucent quartz stones as compared to the ambient light intensity from September 30–October 5, 2004. Fluctuations of the ambient light intensity are closely resembled by light intensities within the hypolithic habitat.

Condensation

Potential condensation conditions occurred almost exclusively from the late evening until the early morning hours (from dawn to dusk). During the time span from May 13 until October 19, 2005, potential water condensation occurred on the quartz surface during 98 of the 160 nights (61%), whereas on the soil surface potential condensation conditions were only present on 74 nights (46%). On some nights, these conditions coincided with rain events. Considering only nights without rain at least 1 h prior to or during condensation conditions, there were still 80 nights (50%) with condensation conditions on quartz, but only 54 nights (34%) with these conditions on the soil crust surface.

On an hourly basis, potential condensation conditions on the surface of the quartz pebbles occurred for 538 h in total, corresponding to

14.0% of the time, whereas on the soil surface comparable conditions occurred for only 355 h or 9.2% of the total time. When condensation conditions within 1 h of the last rain event were eliminated, there were still 498 h (corresponding to 13.0% of the time) on quartz pebbles and 316 h (8.2%) on the soil crust surface.

Soil Moisture

Soil moisture contents were analyzed for a time span from September 1, 2005 to August 31, 2006. During that year, soil moisture at a depth of 30 mm revealed large differences, depending on whether the sensor was placed below the surface of a soil-inhabiting crust or a hypolithic crust (Figure 7). The soil moisture below the soil-inhabiting crust was almost always close to zero from October through April, and only larger rain events, as on October 17, 2005 (5.4 mm) and January 26, 2006 (3.2 mm) led to a short-term increase in soil moisture. Under the quartz surface, soil moisture always remained somewhat higher and the "period of dry soil" only lasted for approximately 6 weeks from the beginning of December to January 21. Smaller rain events during the dry season, as on January 21 (2 mm), February 6 (2.6 mm), and March 2 (1.2 mm), led to much larger soil moisture values under the quartz surface, which lasted for 81, 124, and 106 h, respectively, whereas under the soil crust, such small events only led to a short-term increase in soil moisture or caused no increase at all.

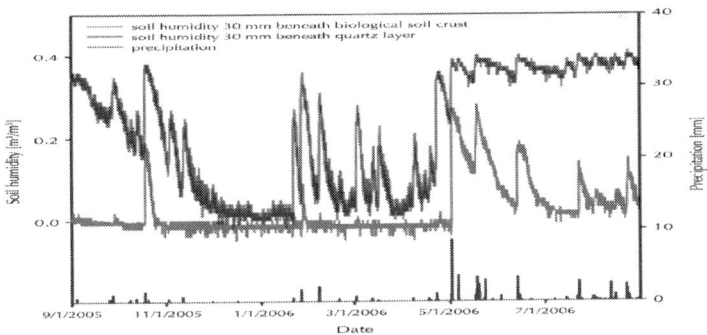

Figure 7: Soil moisture at a depth of 30 mm measured below biological soil crust and below a quartz layer from September 2005 through July 2006. Precipitation values are shown for data interpretation.

An intense rain event at the end of the dry season on May 1 (16.8 mm) caused a substantial increase in soil moisture below both crust types. The water content below the soil-inhabiting crust was still lower than that under the quartz layer and it decreased much quicker. Soil moisture under the quartz layer remained high with values above 0.3 m^3/m^3 almost continuously from April 28 to August 31.

DISCUSSION

Evaluation of the two biological soil crust habitats—soil-inhabiting and hypolithic crusts—revealed obvious microclimatic differences, i.e., moderated temperature variation, lower light intensities, and wetter soil conditions within the latter habitat. Nevertheless, almost identical biomass values were measured in both habitats, suggesting that the two habitats are occupied by stable communities of photoautotrophic organisms. In addition to a set of widely occurring genera, we observed*Chroococcidiopsis* to be restricted to the hypolithic habitat, whereas *Microcoleus* and *Nostoc* occurred almost exclusively in the soil habitat.

Chlorophyll values of 86.1 and 87.7 mg Chl_a/m^2 and 133.7 and 136.2 mg Chl_{a+b}/m^2 measured in the soil and hypolithic habitat, respectively, are well within the range reported by Broady (1981), who observed Chl_a values of 64 mg/m^2 (SD = 40) for sublithic communities within moist mineral soil and 120 mg/m^2 (SD = 60) within dry mineral soil in Antarctica. Büdel et al. (2009), who studied biological soil crusts along a transect in southern Africa, measured very similar values in late successional lichen- and bryophyte-dominated crusts with values of 86.3 mg Chl_a/m^2 and 135.8 mg Chl_{a+b}/m^2, whereas for initial, light cyanobacterial crusts (31.8 mg Chl_a/m^2 and 49.3 mg Chl_{a+b}/m^2) and well-developed cyanobacterial crusts (67.9 mg Chl_a/m^2 and 112.5 mg Chl_{a+b}/m^2) lower biomass values were reported. On a biome-basis, higher biomass values were only reached within the Namib Desert (141.5 mg Chl_a/m^2 and 230.1 mg Chl_{a+b}/m^2), whereas in the Dry Forest, Dry Savanna, Nama Karoo, and Kalahari, values were clearly lower (Büdel et al. 2009).

As a first assumption, one might expect higher biomass values in the hypolithic environment, and we could imagine two potential causes for our equal values. First, the hypolithic habitat is limited in space, i.e., all favorable stones are fully colonized, and biomass could reach higher values if more suitable hypolithic habitat was available (e.g., at higher quartz stone density). Second, when activated by water, the soil-inhabiting crusts reach a far higher net photosynthesis rate due to the higher light intensities at the soil surface. Unfortunately, CO_2 gas exchange measurements do not serve as a reliable method here, since bare Knersvlakte soils have been observed to release CO_2.

An estimation of biomass values by means of chlorophyll values has to be considered with care, since chlorophyll contents of autotrophic organisms are known to vary, e.g., with changing nutrient status of the organisms (Kruskopf and Flynn 2006). Nevertheless, chlorophyll determination represents the most appropriate method to assess the biomass of the photosynthetically active parts of BSC. Since nutrient distributions within our study area are considered to be fairly homogeneous, this method is estimated to be well suited to meet our task.

For the coverage of quartz stones, almost identical values were determined by Vogel (1955), who found 43% versus 46% determined by us. The frequency of hypolithic photoautotrophs has been determined in a variety of hot and cold deserts to range from 0 to 100% (Table 3). Annual precipitation seems to be one factor influencing the frequency of hypoliths and rock type also has an impact, as Berner and Evenari (1978) showed with different colonization frequencies on light and dark flint in the Negev Desert. On the mainland and islands of Baja California, frequency of hypoliths on quartz pebbles reached about half the values observed by us at about half the amount of annual precipitation (Heckman et al. 2006). In cold deserts, large ranges in the colonization rates were observed depending on the impact of fog (Azúa-Bustos et al. 2011) and snowmelt (Pointing et al. 2009) in addition to the precipitation influence (Azúa-Bustos et al. 2011). With decreasing precipitation and fog, frequency approached zero (Warren-Rhodes et al. 2006). Extraordinarily high frequencies of 95 and 94% were documented by Cockell and Stokes (2006) on Devon and Cornwallis Island, respectively. Schlesinger et al. (2003) even reported that 100% of all quartz pebbles in the southern Mojave Desert are colonized by photoautotrophs, despite the fact that the annual precipitation rate is even lower than in our study area. It also is striking that the smallest pebbles they found had a thickness of 5 mm and were all colonized, whereas we found only 26% of the pebbles between 4 and 5 mm to be colonized by cyanobacteria.

Table 3: Frequency of hypolithic photoautotrophs within different hot and cold deserts

Location	Frequency of hypolithic photoautotrophs (%)	Rock type	Annual precipitation (mm)	Reference
Hot deserts				
Mojave Desert, USA	100	Quartz	83	Schlesinger et al. (2003)
Knersvlakte, South Africa	69	Quartz	126	This study
Negev Desert, Israel	46.8	Dark flint	76	Berner and Evenari (1978)
Mainland of Baja California, Mexico	38	Quartz	53	Heckman et al. 2006
Islands of Baja California, Mexico	26	Quartz	53	Heckman et al. 2006
Negev Desert, Israel	20.9	Light flint	76	Berner and Evenari (1978)
Cold deserts				
Alexander Island, Antarctica	86–100	"Opaque rocks"	-	Cockell and Stokes 2006
Devon and Cornwallis Island, Canada	95 and 94	"Opaque rocks"	-	Cockell and Stokes 2006
Atacama Desert, Chile	80	Quartz	- (Fog input)	Azúa-Bustos et al. 2011
High altitude tundra, central Tibet	36	Quartz	-	Wong et al. 2010
Copiapó, Atacama Desert, Chile	27.6	Quartz and quartzite	~ 110	Warren-Rhodes et al.2006
McMurdo Dry Valleys, Antarctica	22	Quartz	- (Snowmelt influence)	Pointing et al. 2009
McMurdo Dry Valleys, Antarctica	4.9	Quartz	-	Pointing et al. 2009
Yungay, Atacama Desert, Chile	0	Quartz and quartzite	~ 5	Warren-Rhodes et al.2006

Additional information is given on the colonized rock type, the annual precipitation within the region (as far as available), and the reference of the data.

Weber et al.

Weber et al. Ecological Processes 2013 2:8, doi:10.1186/2192-1709-2-8

We observed the colonized stones to be significantly thicker (mean thickness: 9.5 mm) than uncolonized ones (mean thickness: 5.7 mm), which is in accordance with Azúa-Bustos et al. (2011), who also found increasing colonization rates with increasing quartz rock size. In contrast, Berner and Evenari (1978) reported larger mean stone thickness values for uncolonized light and dark flint stones (16–20 mm) as compared to colonized ones (11–15 mm).

Our data revealed colonized quartz stones to be significantly heavier (mean value: 3.4 g) than uncolonized ones (mean value: 0.7 g). Despite the fact that the mass of stones is an essential factor influencing stability of the hypolithic habitat, no comparable data could be found in the literature.

Unfortunately, our microclimate sensors were not installed in sufficient numbers to allow for statistical analyses. Nevertheless, we believe that our data may serve as an important resource, as we are the first to describe the long-term microclimatic conditions within the soil crust and hypolithic habitat in the Knersvlakte.

Our measurements clearly revealed that temperature conditions under the quartz stones were less extreme compared to the quartz stone and soil surface. This finding was also reported by Azúa-Bustos et al. (2011) for the Atacama Desert and is generally expected for both hot and cold deserts (Chan et al. 2012). However, Schlesinger et al. (2003) measured larger temperature amplitudes and higher temperature maxima under the quartz stones as compared to the stone surface. They describe this phenomenon, which is puzzling to us and contradictory to all the other data, as a modest greenhouse effect under the quartz stones. Absolute temperatures measured by us in the hypolithic environment ranged from 4.2 to 59.7°C, which is surprisingly similar to the values observed by Berner and Evenari (1978) in the Negev Desert (about 2 to 50°C) and by Tracy et al. (2010) in northern Australia (about 9 to 60°C).

We observed light intensities under the quartz rocks to be clearly lower, but still reaching about 15–30% of the ambient light intensities, which were closely tracked. This is somewhat higher than the measurements of Vogel (1955), who found a transmission of about 9.4% of the ambient light at a stone thickness of 10 mm. Our higher values may be caused by a somewhat shallower embedding of the quartz pebbles after installation of the light sensors as compared to their original position in the soil. On the other hand, Vogel (1955) described his lab measurements as an estimation that did not account for scattered light coming in from the sides, which may again increase the amount of incoming light. In contrast to that, Berner and Evenari (1978) measured only 0.005% of the incident light at a depth of 10 mm, Schlesinger et al. (2003) observed 2.8% of the incident light at a depth of 12 mm (wavelength: 680 nm), and Broady (1981) found that 2.7% of the incident light reached the bottom of a 13 mm thick stone. A major drawback of all these measurements (including ours) is that only light descending from the upper part of the rock was measured. Due to light scattering and reflectance within the stone and the surrounding soil layer, light most probably also reaches the organisms from the sides and even from underneath. In order to measure this, especially adapted scalar irradiance sensors have to be applied (Kühl et al. 1994).

We calculated condensation to occur much more frequently on the quartz stone surface as compared to the soil surface (538 vs. 355 h corresponding to 14.0 vs. 9.2% of the time from May 13 to October 19, 2005). If the influence of precipitation, which may cool down the quartz or soil surface, was excluded, there were still 498 vs. 316 h of condensation conditions. Condensation conditions without rain influence were observed during 50% of the nights on quartz pebble surfaces, but only 34% of the nights on soil crust surfaces. This difference is most probably caused by the higher heat transfer rates of quartz stones as compared to the soil surface, causing the quartz pebbles to cool down more quickly. The condensed water trickles along the quartz stone sides and infiltrates deeper into the soil, as Vogel (1955) observed in an experimental approach. During drying he found that, due to the blocked evaporation, the soil under the stone remained wet for a longer time. This perfectly agrees with our measurements of far higher and also more balanced humidity values under the quartz field as compared to the soil-inhabiting crust. The lower maximum soil moisture may also be caused by a lower pore space of soil below soil-inhabiting crusts as compared to hypolithic crusts. Higher moisture contents under the quartz stones as compared to the surrounding soil were also found by Smith et al. (2000), and in the absence of rain, Azúa-Bustos et al. (2011) observed humidity values above 90% in the early morning hours.

CONCLUSIONS

Our measurements revealed that, in spite of large microclimatic differences, almost identical biomass values were reached by the soil-inhabiting and hypolithic communities. Whereas the soil habitat was characterized by more extreme temperatures, short infrequent precipitation-, fog-, and dewfall events, and high light intensities, the temperatures in the hypolithic environment were more balanced and water conditions were much more favorable, while light intensities were clearly lower.

Microscopical studies revealed that *Chroococcidiopsis* only occurred in the hypolithic habitat, whereas *Microcoleus* and *Nostoc* were almost exclusively observed in the soil habitat. In further studies it would be interesting to investigate the physiological preferences, adaptation, and acclimatization of these communities.

AUTHORS' CONTRIBUTIONS

BW, KD, SD, DW, and BB designed the project; DW designed the climate station setup; SD and BB conducted taxonomic determinations; BW, DW, KD, HR, and BB established the climate station in the field; BW, DW, KD and BB evaluated quartz stone colonization; DW KD and BW analyzed the field data; and DW and BW wrote the manuscript. All authors carefully read and approved the final manuscript.

ACKNOWLEDGEMENTS

Various forms of assistance from the Universities of Limpopo (South Africa) and Kaiserslautern (Germany) are gratefully acknowledged. We would like to thank Douw Venter, Rudi Wetterman, and Martin

Potgieter (University of Limpopo) for their help during the design and installation of the climate stations. Research in the Knersvlakte, South Africa, was conducted with Northern Cape research permits (No. 10/2005, ODB 052/2006) and the appending export permits. This project was financed by the German Ministry for Education and Research (BmBF) in the framework of BIOTA South (promotion number 01 LC 0024A).

REFERENCES

1. Arnon DI, McSwain BD, Tsuijmoto HY, Wada K (1974) Photochemical activity and components of membrane preparations from blue-green algae. I. Coexistence of two photosystems in relation to chlorophyll a and removal of phycocyanin. Biochim Biophys Acta 357:231-245

2. Azúa-Bustos A, González-Silva C, Mancilla RA, Salas L, Gómez-Silva B, McKay CP, Vicuña R (2011) Hypolithic cyanobacteria supported mainly by fog in the Coastal Range of the Atacama Desert. Microb Ecol 61:568-581

3. Belnap J, Büdel B, Lange OL (2003) Biological soil crusts: characteristics and distribution. In: Belnap J, Lange OL (eds) Ecological studies 150: biological soil crusts: structure, function, and management, Berlin, Heidelberg, New York: Springer. pp 3-30

4. Berner T, Evenari M (1978) The influence of temperature and light penetration on the abundance of the hypolithic algae in the Negev Desert of Israel. Oecologia 33:255-260

5. Bhatnagar A, Bhatnagar M (2005) Microbial diversity in desert ecosystems. Curr Sci India 1:91-100

6. Broady PA (1981) The ecology of sublithic terrestrial algae at the Vestfold Hills, Antarctica. Brit Phycol J 16:231-240

7. Büdel B, Schultz M (2003) A way to cope with high irradiance and drought: inverted morphology of a new cyanobacterial lichen, Peltula inversa sp. nov., from the Nama Karoo, Namibia. Bibl Lichenol 86:225-232

8. Büdel B, Wessels DCJ (1991) Rock inhabiting blue-green algae/cyanobacteria from hot arid regions. Arch Hydrobiol Suppl Algol Stud 64:385-398

9. Büdel B, Darienko T, Deutschewitz K, Dojani S, Friedl T, Mohr K, Salisch M, Reisser W, Weber B (2009) Southern African biological soil crusts are ubiquitous and highly diverse in drylands, being restricted by rainfall frequency. Microb Ecol 57(2):229-247

10. Cameron RE (1960) Communities of soil algae occurring in the Sonoran desert in Arizona. J Ariz-Nev Acad Sci 1:85-88

11. Cameron RE, Blank GB (1967) Soil studies: microflora of desert regions, VIII. Distribution and abundance of desert microflora. JPL Space Programs Summary No 37–44 4:193-201 Pasadena, CA

12. Chan Y, Lacap DC, Lau MCY, Ha KY, Warren-Rhodes KA, Cockell CS, Cowan DA, McKay CP, Pointing SB (2012) Hypolithic microbial communities: between a rock and a hard place. Environ Microbiol 14(9):2272-2282

13. Cockell CS, Stokes MD (2004) Widespread colonization by polar hypoliths. Nature 431:414

14. Cockell CS, Stokes MD (2006) Hypolithic colonization of opaque rocks in the Arctic and Antarctic Polar Desert. Arct Antarct Alp Res 38(3):335-342

15. Cowan DA, Khan N, Pointing SB, Cary SC (2010) Diverse hypolithic refuge communities in the McMurdo Dry Valleys. Antarct Sci 22(6):714-720

16. Cowling M, Esler KJ, Rundel PW (1999) Namaqualand, South Africa—an overview of a unique winter-rainfall desert ecosystem. Plant Ecol 142:3-21

17. Friedmann EI, Galun M (1974) Desert algae, lichens, and fungi. In: Brown GW (ed) Desert biology, vol II, London: Academic. pp 165-212

18. Friedmann EI, Lipkin Y, Ocampo-Paus R (1967) Desert algae of the Negev (Israel). Phycologia 6:185-196

19. Galun M, Garty J (2003) Biological soil crusts of the Middle East. In: Belnap J, Lange OL (eds) Biological soil crusts: structure, function, and management, Berlin, Heidelberg, New York: Springer. pp 95-106

20. Geitler L (1932) Cyanophyceae von Europa unter Berücksichtigung der anderen Kontinente. Leipzig: Akademische Verlagsgesellschaft.

21. Heckman KA, Anderson WB, Wait DA (2006) Distribution and activity of hypolithic soil crusts in a hyperarid desert (Baja California, Mexico). Biol Fertil Soils 43:263-266

22. Jürgens N, Haarmeyer DH, Luther-Mosebach J, Dengler J, Finckh M, Schmiedel U (2010) Biodiversity in southern Africa. Volume 1: Patterns at local scale—the BIOTA Observatories. Klaus Hess, Göttingen.

23. Komárek J, Anagnostidis K (1998) Cyanoprocaryota 1. Gustav Fischer, Jena: Teil Chroococcales.

24. Komárek J, Anagnostidis K (2005) Cyanoprocaryota 2. Elsevier, München: Teil Oscillatoriales.

25. Kühl M, Lassen C, Jørgensen BB (1994) Light penetration and light intensity in sandy marine sediments measured with irradiance and scalar irradiance fiber-optic microprobes. Mar Ecol Prog Ser 105:139-148

26. Kruskopf M, Flynn KJ (2006) Chlorophyll content and fluorescence responses cannot be used to gauge reliably phytoplankton biomass, nutrient status or growth rate. New Phytol 169:525-536

27. Lacap DC, Warren-Rhodes KA, McKay CP, Pointing SB (2011) Cyanobacteria and chloroflexi-dominated hypolithic colonization of quartz at the hyper-arid core of the Atacama Desert, Chile. Extremophiles 15:31-38

28. Müller G (2009) Hypolithic plants from Carruthers Peak, Snowy Mountains, New South Wales, Australia. Geogr Res 47(4):449-453

29. Pointing SB, Warren-Rhodes KA, Lacap DC, Rhodes KL, McKay CP (2007) Hypolithic community shifts occur as a result of liquid water availability along environmental gradients in China's hot and cold hyperarid deserts. Environ Microbiol 9(2):414-424

30. Pointing SB, Chan Y, Lacap DC, Lau MCY, Jurgens JA, Farrell RL (2009) Highly specialized microbial diversity in hyper-arid polar desert. Proc Natl Acad Sci USA 106:19964-19969

31. Ronen R, Galun M (1984) Pigment extraction from lichens with dimethyl sulfoxide (DMSO) and estimation of chlorophyll degradation. Environ Exp Bot 24(3):239-245

32. Rummrich U, Rummrich M, Lange-Bertalot H (1989) Diatomeen als "Fensteralgen" in der Namib-Wüste und anderen ariden Gebieten von SWA/Namibia. Dinteria 20:23-29

33. Schlesinger WH, Pippen JS, Wallenstein MD, Hofmockel KS, Klepeis DM, Mahall BE (2003) Community composition and photosynthesis by photoautotrophs under quartz stones, southern Mojave Desert. Ecology 84:3222-3231

34. Schmiedel U (2002) The quartz fields of southern Africa. Flora, phytogeography, vegetation and habitat ecology. Cologne: PhD Dissertation, University of Cologne.

35. Smith MC, Bowman JP, Scott FJ, Line MA (2000) Sublithic bacteria associated with Antarctic quartz stones. Antarct Sci 12(2):177-184

36. Tracy CR, Streten-Joyce C, Dalton R, Nussear KE, Gibb KS, Christian KA (2010) Microclimate and limits to photosynthesis in a diverse community of hypolithic cyanobacteria in northern Australia. Environ Microbiol 12(3):592-607

37. Ullmann I, Büdel B (2003) Biological soil crusts of Africa. In: Belnap J, Lange OL (eds) Biological soil crusts: structure, function, and management, Berlin, Heidelberg, New York: Springer. pp 107-118

38. Vogel S (1955) Niedere "Fensterpflanzen" in der südafrikanischen Wüste. Eine ökologische Sondierung. Beitr Biol Pflanz 31:45-135

39. Warren-Rhodes KA, Rhodes KL, Pointing SB, Ewing SA, Lacap DC, Gómez-Silva B, Amundson R, Friedmann EI, McKay CP (2006) Hypolithic bacteria, dry limit of photosynthesis and microbial ecology in the hyperarid Atacama Desert. Microb Ecol 52:389-398

40. Weber B, Deutschewitz K, Dojani S, Friedl T, Darienko T, Mohr K, Büdel B (2010) Goedehoop (S26): Biological soil crusts. In: Jürgens N, Haarmeyer DH, Luther-Mosebach J, Dengler J, Finckh M, Schmiedel U (eds) Biodiversity in southern Africa. Volume 1:

Patterns at local scale—the BIOTA Observatories, Klaus Hess, Göttingen.

41. Wong FKY, Lacap DC, Lau MCY, Aitchison JC, Cowan DA, Pointing SB (2010) Hypolithic microbial community of quartz pavement in the high-altitude tundra of central Tibet. Microb Ecol 60:730-739

Citations

CHAPTER 1

Drewniak Lukasz, Rajpert Liwia, Mantur Aleksandra, and Sklodowska Aleksandra, "Dissolution of Arsenic Minerals Mediated by Dissimilatory Arsenate Reducing Bacteria: Estimation of the Physiological Potential for Arsenic Mobilization," BioMed Research International, vol. 2014, Article ID 841892, 12 pages, 2014. doi:10.1155/2014/841892.

CHAPTER 2

Gui Jin, Zhaohua Li, Qiaowen Lin, Chenchen Shi, Bing Liu, and Lina Yao, "Land Use Suitability Assessment in Low-Slope Hilly Regions under the Impact of Urbanization in Yunnan, China," Advances in Meteorology, Article ID 848795, in press.

CHAPTER 3

Meirui Zhong, Anqi Zeng, Jianbai Huang, and Jinyu Chen, "The Analysis of Pricing Power of Preponderant Metal Mineral Resources under the Perspective of Intergenerational Equity and Social Preferences: An Analytical Framework Based on Cournot Equilibrium Model," Abstract and Applied Analysis, vol. 2014, Article ID 252739, 11 pages, 2014. doi:10.1155/2014/252739.

CHAPTER 4

Felix Oteng Mensah, Clement Alo, and Sandow Mark Yidana, "Evaluation of Groundwater Recharge Estimates in a Partially Metamorphosed Sedimentary Basin in a Tropical Environment: Application of Natural Tracers,"The Scientific World Journal, vol. 2014, Article ID 419508, 8 pages, 2014. doi:10.1155/2014/419508.

CHAPTER 5

Benoit Lafleur, Nicole J Fenton, and Yves Bergeron, Forecasting the Development of Boreal Paludified Forests in Response to Climate Change: a Case Study using Ontario Ecosite Classification, doi:10.1186/s40663-015-0027-6.

CHAPTER 6

Vanesa Rodríguez Osuna, Jan Börner, Udo Nehren, Rachel Bardy Prado, Hartmut Gaese, and Jürgen Heinrich, Priority Areas for Watershed Service Conservation in the Guapi-Macacu Region of Rio de Janeiro, Atlantic Forest, Brazil, doi:10.1186/s13717-014-0016-7.

CHAPTER 7

Axel Bergmann, Frank-Andreas Weber, H Georg Meiners, and Frank Müller, Potential Water-related Environmental Risks of Hydraulic Fracturing Employed in Exploration and Exploitation of Unconventional Natural Gas Reservoirs in Germany, doi:10.1186/2190-4715-26-10.

CHAPTER 8

Bettina Weber, Dirk CJ Wessels, Kirstin Deutschewitz, Stephanie Dojani, Hans Reichenberger, and Burkhard Büdel, Ecological Characterization of Soil-inhabiting and Hypolithic Soil Crusts within The Knersvlakte, South Africa, doi:10.1186/2192-1709-2-8.

Index

A

Areas of permanent protection (APP) 161
Arsenic 27, 28, 31

B

Biological soil crusts (BSC) 227
Biotechnological tool 11
Brazilian Atlantic Forest Trust (BART) 157

C

Chemical Abstracts Service (CAS) 207
Chemical Oxygen Demand (COD) 177
Chemolithoautotrophic microorganism 20
Chloride mass balance (CMB) 96, 97, 110
Coalbed methane (CBM) 195
Colony forming unit (CFU) 8
Community Water and Sanitation Agency (CWSA) 101

D

Digital elevation model (DEM) 38
Dissimilatory arsenate reducing bacteria (DARB) 9
Driving Forces, Pressure, State, Impact and Response (DPSIR) 161
Dynamic factor 125
Dynamic process 116, 117

E

Environmental impact assessment (EIA) 199
Extension and Technical Assistance Agency (EMATER) 159

F

Farming systems (FS) 165
Flame atomic absorption spectrometry (FAAS) 6
Food and Agriculture Organization (FAO) 35

G

Gas in place (GIP) 195
Global Environmental Facility (GEF) 154
Global meteoric water line (GMWL) 104

H

High pressure 35
Human geography 56

L

Local Groundwater Line (LGWL) 105
Local Meteoric Water Line (LMWL) 105
Lysogeny broth (LB) 4

M

Mechanisms of microbial arsenic 13
Microbial mat 4, 9
Microclimatic thermal 233
Minimal inhibitory concentrations (MIC) 7

N

Nitrogen (N) 162
NORM (Naturally Occurring Radioactive Material) 210

O

Opportunity costs (OCs) 153, 155, 159

P

Payments for ecosystem services (PES) 146
Payments for watershed services (PWS) 146
Phosphorous (P) 162
Piracicaba, Capivari and Jundiaí (PCJ) 154
Preponderant metal mineral 76
Pricing power 62

S

Soil organic layer (SOL) 130, 131

T

Territory 117, 119, 129, 133, 134, 136

V

Vienna Standard Mean Ocean Water (VSMOW) 102

W

Water Management Act (WHG) 199
Willingness to pay (WTP) 163
World Bank (WB) 154